Digital, and its Value

ISBN: 978-1-7374446-1-9 (Paperback)
ISBN: 978-1-7374446-2-6 (Hardcover)
ISBN: 978-1-7374446-0-2 (eBook)

Library of Congress Control Number: 2021916103

Limit of Liability/Disclaimer of Warranty: This book is based on the author's learnings, experience, and research. While the author has used his best efforts to research and prepare this book, the author assumes no responsibility for errors or for the accuracy or completeness of the contents of this book. The author disclaims any implied warranties of any type. This book is sold on the understanding that under no circumstance does the author assume liability of any type resulting from the use of information in this book or otherwise, and that the author shall not be liable for damages of any type from the use of this book and its contents. This book and its contents are meant to inform the reader based on the author's learnings and experience; nothing in this book should replace professional or legal advice or common sense.

Any trademarks, company names, service names, brand names, product names, or any other named features used in this book are used only for reference and are assumed to be the property of their respective owners. References to websites (URLs) were accurate at the time of writing this book. The author is not responsible for URLs or their content that may have changed or expired since the time of preparation of the manuscript.

Front cover design by Shelley Savoy
Book design by Booknook.biz

Printed in the United States of America.
First printing edition 2021.

Digital, and its Value

What exactly is digital and how does it create value.

A primer for C-suites, executives,
and MBA students

by

Vishal Saxena

Dedicated to my parents, whose sacrifices have created all my possibilities.

Contents

Introduction

The word "Digital" is everywhere. Matching in popularity is its conjoint twin: the word "Transformation".

But digital as a term can be associated with computers for as long as they have been around. After all, the digits '1' and '0' have represented the binary language that machines use to execute human ordained computer instructions, for decades. Thus, at least in this sense, the word "Digital" is not new.

Similarly, the business processes have been getting transformed since the commercial advent of the internet in the '90s, when the manual business operations started to get digitized earnestly into systems such as for enterprise resource management, customer relationship management, and business intelligence.

So, if digital has existed for decades, as also transformation, then what is the new thing about the digital of today? Is it more a case of old stuff, perhaps embellished a little, but then repackaged as new? Or, is there something more profound and more urgent to it?

The answer unequivocally and cutting to the chase is: Digital today needs to be seen as an existential proposition for companies. Ignore it at your company's peril.

This book starts by briefly substantiating this argument.

Then, the book lays out the market and technology forces that act as the demand patterns for companies to embrace digital in its contemporary avatar. These patterns are the reasons companies need to embrace digital. Embraced well, digital can help the companies grow both top-line revenues and reduce costs, leapfrogging their competition. Most leading companies of the world today understand this. And it is no surprise that there are not many business strategies worth their salt these days that are not hinged on digital.

Even so, a good understanding of the words "digital" and "transformation" is lacking. What is missing is a good grasp of the answers to two key questions: "What exactly is Digital?" and "How does it create value?"

Perhaps as symptoms of this lack of understanding are the common questions that get asked by the C-Suites, and executives charged with driving top-line growth and margin improvement. These questions include: "Where do we start?", "What digital investments should we fund today to sustain our competitive advantages of today and, to create advantages for our future?", "What digital capabilities are we missing and must have?"

In my interactions with C-Suites and line executives of the companies that I have served as Management Consultant over the years, I have found that a missing genuine appreciation of what digital is, the opportunities, and the threats it represents, often leads to nervousness in strategic planning meetings. Worse, the same lack of understanding is often prevalent in the people who are advising the companies on all things digital. Frequently, these service-providing leaders have an appreciation of only one or two aspects of digital and are missing the big picture view of all the value levers that digital can—and ought to—drive.

To meet this need, this book has been written for C-Suites and line executives who are chartered by their Boards to make full use of digital. Also, part of the target audience is the service-providing executive who is dealing with digital in either a selling or a delivery role. Technology or human resources or other functional executives who have morphed into doing things digital, or hiring for digital, are targeted core audiences as well. Finally, most business school curriculums do not cover digital well, or at least to the extent that they should. To that end, MBA students looking to, well, understand what exactly digital is and how does it create value should benefit from this book as well.

This book is divided into four parts.

The first part makes a brief argument for why digital needs to be viewed as having an existential relevance for modern corporations.

The second part reviews the demand side patterns for digital. As just discussed, these patterns represent the market and technology innovations or developments of recent years that have shaped the definition of digital and its value drivers. Patterns covered in this section can be seen as the reasons why organizations of all hues, shapes, and sizes across the globe need to respond to digital.

The third part covers the core levers that the organizations are using to create value from digital—in response to the demand patterns. These levers that in addition to technology, also cover the modern organization and management models, comprise the digital palette that is used to create colors of value. This is the section that answers the question: What exactly is digital? Also presented in this section is a Digital Capability Model that companies can use to identify, prioritize, and sequence their investments for digital.

The final and fourth part now looks at how the value from digital is created using the building blocks presented in the previous section. We take two views of value. The first view takes a look at how value can be created from digital in a corporate function. The functions covered are Inventory Management; Sales, Marketing and Customer Service; Finance, and HR. The second view of value considers value creation from digital, by industry type. The selected assortment of industry sectors covers Retail; Pharmaceutical; Manufacturing; and Banking. This assortment should give the reader a good flavor of how value is created from digital in asset-intensive industries such as Manufacturing; heavily regulated industries such as Pharmaceutical and Banking; and less regulated and relatively less asset-intensive industries such as Retail.

It is hoped that through this book, C-Suites and business strategists can design, mobilize and charge their viable business strategies using digital.

Part One

The Existential Argument

Et Tu, Digital?

Saddled with debt from leveraged buyouts, J. Crew and Neiman Marcus—the two well-known retail brands—paid their private equity owners hundreds of millions of dollars in interest and fees rather than making investments to address the needs of a transforming retail environment[1]. Even before the pandemic hit, both the brands had trouble adjusting well to the rising trend of e-commerce and to connect with a "new generation" of shoppers[2]. What the pandemic did was perhaps shorten the path to an eventual crisis—when the pandemic eliminated much of their sales, both the brands had to seek relief from court[3]. J. Crew filed for bankruptcy on May 4, 2020[4, 5]. Neiman Marcus filed for bankruptcy on May 7—three days later[6].

Changed shopping behavior of customers is a major contributor to the transformed retail environment.

But what are the contributing factors to this changing shopping behavior? Is it the same customer segments that are behaving differently in how they shop or are there new customer segments that have emerged in the market?

The answer is the latter.

There have been two new consumer demographics that have emerged whose preferences for how they interact with a brand are markedly dif-

---

1. Maheshwari, S., & Friedman, V. (2020, May 14). "The Pandemic Helped Topple Two Retailers. So Did Private Equity." *The New York Times*. https://www.nytimes.com/2020/05/14/business/coronavirus-retail-bankruptcies-private-equity.html.
2. Ibid.
3. Ibid.
4. www.prnewswire.com/news-releases/jcrew-group-inc-announces-comprehensive-agreement-to-deleverage-balance-sheet-and-position-jcrew-and-madewell-for-long-term-profitable-growth-301051688.html. Accessed 28 July 2021.
5. Wikipedia. en.wikipedia.org/wiki/J.Crew. Accessed 28 July 2021.
6. wwd.com/fashion-news/fashion-scoops/neiman-marcus-bankruptcy-what-to-know-1203565951. Accessed 28 July 2021.

ferent from those of previous generations. The economic clout and the sheer numbers that these two new consumer segments represent make them a force to be reckoned with for the brands. Together, these segments exercise an outsized influence on the viability of brands.

The first of these is the Gen Y or the Millennial consumer segment that comprises about 72 million of the US population and consists of people who were born between the years 1981 and 1996[7]. The second segment is Gen Z, comprising of people who were born between 1997 and 2012 and making up about 68 million of the US population[8]. Together, the Millennials and the Gen Z account for about 42 percent of the US population.

These two youth segments—the Gen Y and the Gen Z—grew up with a strong disposition for technology. The Millennial or the Gen Y segment grew up as the adoption of the personal computer was increasing. In 1989, Berners-Lee wrote the proposal for the World Wide Web, and by the end of 1990, the first web page arrived on the open internet[9, 10]. It is safe to say that while Gen Y saw the period that led to the birth of the world wide web, Gen Z grew up after its birth.

No wonder then that both Gen Y and Gen Z expect technology to be a major component of how they interact with the brands and what they expect from their brand experiences.

Savvy users of the internet, and now also mobile, these youth generations use technology heavily to discover brands, their products, services, value propositions, and price. They then seek whichever channel—a physical store, or web, or mobile—offers the best value for their money[11]. The industry segments that have the potential to get signifi-

---

7. Kasasa. "Boomers, Gen X, Gen Y, Gen Z, and Gen A Explained." Kasasa, www.kasasa.com/articles/generations/gen-x-gen-y-gen-z. Accessed 29 July 2021.
8. Ibid.
9. "Where the Web Was Born." CERN, 14 Apr. 2021, home.cern/science/computing/where-web-was-born.
10. "History of the Web." World Wide Web Foundation, webfoundation.org/about/vision/history-of-the-web. Accessed 16 June 2021.
11. Morgan Stanley. "How a 'Youth Boom' Could Shake Up Spending Trends." Morgan Stanley, 2019, www.morganstanley.com/ideas/gen-z-millennials-set-for-consumer-spending-increases?ET_MID=noloc&cid=whmt-smfa-howayo-3273. Accessed 16 June 2021.

cantly impacted by this trend in buyer preference include retail, food and beverage, electronics, and travel[12]. But it is not only these sectors that need to respond to a technology-hinged set of customer expectations.

While the baby boomers—those born between 1946 and 1964, and predecessors of Gen Y and Gen Z—preferred to go into a brick and mortar branch to perform their bank transactions, both Millennials and Gen Z show a preference for the digital (versus the physical store) channel[13]. By one estimate, over 50% of Gen Z have not entered a bank branch for at least three months[14].

A typical Gen Z person received her first mobile phone at 10.3 years of age, and many of the Gen Zs grew up playing with their parents' tablets or mobile phones[15]. On average, a Gen Z spends about 3 hours a day on her mobile device[16].

24/7 access to a mobile device and social media means that these technology sophisticates do not hesitate to speak to their friends and network on social media about their experiences with a brand. As a result, a brand's reputation and equity can be built or destroyed at unprecedented speeds.

There is reason to believe that having seen the economic downturns such as from the 9/11 attacks and the 2008 financial crisis, these young consumer segments prize value over loyalty to a brand, with the result that they do not hesitate to switch brands if, for instance, they had a poor customer service experience[17].

This lack of commitment to a brand means that a brand has to be always on its toes to serve these segments well. Even that may not be enough. Rather than a reactive or even a co-active approach (where the brand is learning of the consumer behavior while the consumer is interacting with the brand), the brands have to take a proactive

12. Ibid.
13. Kasasa. "Boomers, Gen X, Gen Y, and Gen Z Explained." Kasasa, www.kasasa.com/articles/generations/gen-x-gen-y-gen-z. Accessed 16 June 2021.
14. Ibid.
15. Ibid.
16. Ibid.
17. Ibid.

approach where they have to systemically identify and respond to consumer needs ahead of the next touchpoint.

As additional challenges for the brand, marketing to these segments comes with its own complexities. Both Gen Z and Gen Y have short attention spans. Gen Z has an attention span of about 8 seconds, while that of Gen Y or Millennials is expected to be about 12 seconds[18]. What this implies is that to market effectively to these segments, the targeted content has to be captivating and very relevant to the consumer's needs and buying preferences[19]. Further, and perhaps as a related effect, these segments show a high tendency to move between devices. While the Millennial can bounce between three screens at a time, Gen Z can be using as many as five screens at the same time[20]. This makes it hard for marketers to reach their targeted audience on the right channel, at the right time, and with the right content.

Ignore these finicky, device-changing, brand-switching, price-sensitive consumer segments, and you could very well be putting your company on a path to declining revenues and eventual bankruptcy.

In addition to comprising substantial chunks of the population, these segments also carry financial clout. Gen Z alone is expected to account for about 40 percent of all consumer markets, in the US, by 2020[21]. Gen Y consumers are expected to earn 46% of US income by 2025[22]. By another measure, 81% of Millennials have employment in the US, with average monthly take-home pay of $3,100, and together with Gen Z teenagers, their combined spending power is estimated to be around $3 Trillion in 2020[23].

Whichever way you look at it, few things are clear. One, the financial clout and the number footprint of the Gen Y and Gen Z consumer

18. Law, Thomas. "10 Vital Strategies to Use When Marketing to Generation Z in 2020." *Oberlo*, 7 Jan. 2021, www.oberlo.com/blog/marketing-strategies-generation-z.
19. Ibid.
20. Ibid.
21. Ibid.
22. Kasasa. "Boomers, Gen X, Gen Y, and Gen Z Explained." Ibid.
23. "Millennials & Gen Z Teens' Combined Spending Power Is Nearly $3 Trillion in 2020." *YPulse*, 15 Jan. 2020, www.ypulse.com/article/2020/01/09/millennials-gen-z-teens-combined-spending-power-is-nearly-3-trillion-in-2020.

segments are expected to become stronger with time. Two, the buying behavior and the expectations that a consumer has of a brand are noticeably different from these two segments than those of prior consumer segments such as the baby boomers. The Gen Y and Gen Z show a strong bias for the digital channel (web or online) than the physical channel (brick and mortar store). Three, the attention spans of these segments—which are short—imply that marketing to these segments requires higher degrees of customer specificity and content attractiveness which suggests a stronger reliance on consumer and channel data.

The retail bankruptcy filings of 2020 informed us how companies' choices related to how the digital ecosystem is used to drive a superior customer experience can lead to an existential crisis.

The shift from physical to digital is unambiguous. And the advent of the pandemic may have accelerated the pace of the transition.

In the same quarter that J. Crew and Neiman Marcus filed for bankruptcy—second quarter of 2020—Walmart and Target reported record sales[24]. Walmart doubled its e-commerce sales; Home Depot's sales increased more than 23 percent in the same quarter, and Wayfair—an online furniture retailer—reported an approximately 80 percent revenue increase year over year[25]. In the first half of 2020, Target added 10 million new digital customers[26]. The strength and convenience of these retailers' e-commerce operations contributed to their success. Their e-commerce platforms offered the consumers a "one-stop shopping advantage" that allowed the consumers to buy everything from groceries to electronics and clothing from the same channel[27]. In the earnings call, Brian Cornell, Target's chief executive, said: "Throughout the crisis, we have deepened our relationship with American consumers and introduced millions of them to our digital fulfillment services."[28].

---

24. Friedman, Gillian. "Walmart, Target and Other Big-Box Retailers See Profits Surge During Pandemic." *The New York Times*, 20 Aug. 2020, www.nytimes.com/2020/08/19/business/coronavirus-walmart-target-home-depot.html.
25. Ibid.
26. Ibid.
27. Ibid.
28. Ibid.

Between March and October 2020—a span of only eight months—Dunkin's share price more than doubled as investors noticed the brand's success in building its app and drive-through services[29].

Adobe Analytics, which scans 80 percent of online transactions of top 100 US web retailers, reported that for the 2020 Black Friday, consumers spent $9 billion—a 21.6 percent increase relative to the 2019 Black Friday[30]. Facteus, which monitors credit and debit card payments in the US, found that for the 2020 Black Friday, department stores' in-person sales fell significantly even though their online sales spiked[31]. The same press article that reported these findings indicated that a large portion of consumer spending had shifted from the physical channel to online "long before the pandemic" even though, the article suggested, the pandemic may have accelerated the pace and the degree of the change[32].

In October 2020, two of Macy's stores in Delaware and Colorado went "dark", meaning that instead of serving as sites where customers come to browse their merchandise and shop, these stores were now serving as fulfillment centers where they process customers' online orders and returns[33]. Macy's response was addressing a core customer buying expectation of modern times: Ability to research products on a channel of their choice (physical, mobile, or web); to purchase on a channel of their choice; and to return using a channel of their choice, while demanding faster shipping for free. Increasingly, and in addition, customers also want to have complete visibility into when the order

29. Hirsch, Lauren. "Dunkin' Brands Is in Talks to Sell Itself and Go Private." *The New York Times*, 25 Oct. 2020, www.nytimes.com/2020/10/25/business/dealbook/dunkin-brands-sale-private-equity.html.
30. Corkery, Michael, and Sapna Maheshwari. "Online Shopping Set a Black Friday Record as Consumers Stayed Home." *The New York Times*, 30 Nov. 2020, www.nytimes.com/2020/11/30/business/online-shopping-set-a-black-friday-record-as-consumers-stayed-home.html.
31. Ibid.
32. Ibid.
33. Corkery, Michael, and Sapna Maheshwari. "As Customers Move Online, So Does the Holiday Shopping Season." *The New York Times*, 22 Dec. 2020, www.nytimes.com/2020/11/23/business/retailers-ecommerce-black-friday.html.

was shipped and when it will be delivered. And then, they also expect notification—along with a picture at times, to prevent theft—for when the order was delivered. Finally, they expect to be able to provide quick feedback on the quality, timeliness, and likability of their delivery experience. Of course, customers need to be able to provide feedback on their product or service experience—that goes without saying.

Both J. Crew and Neiman Marcus have since exited their bankruptcies[34, 35]. Post exit, J. Crew's leadership envisions "augmenting the brand experience to improve the retailer's relationship with customers" and leveraging the company's e-commerce platforms "to succeed in today's evolving retail landscape"[36]. Nordstrom, a rival in the US retail market, recently said that it expects that its digital business will make up 50% of its total sales in its fiscal 2021; this number was 33% in 2019, only a couple of years ago[37].

These examples, which map to the retail and consumer industry sectors, might suggest that this digitally-driven consumer tsunami may be restricted to just those and related sectors. To appreciate why this may not be so, let us just remind ourselves of the traits of the modern shopper. While she expects to shop more online, using the channel and time of her choice, she is also sticky with the mobile and the online channel. If she is purchasing her groceries, clothing, and even furniture through these channels, she may also not want: to visit a bank branch to make her deposits and inquire about her loans; to visit a health care provider for her medical needs, and would rather use telemedicine instead; to shop for her car, home and auto insurance

---

34. FashionNetwork.com US. "J. Crew Emerges from Bankruptcy, Outlines New Strategy." FashionNetwork.Com, us.fashionnetwork.com/news/J-crew-emerges-from-bankruptcy-outlines-new-strategy,1242449.html. Accessed 3 July 2021.
35. Reuters. "Neiman Marcus Emerges from Bankruptcy." CNBC, 25 Sept. 2020, www.cnbc.com/2020/09/25/neiman-marcus-emerges-from-bankruptcy.html.
36. FashionNetwork.com US. "J. Crew Emerges from Bankruptcy, Outlines New Strategy." FashionNetwork.Com, ibid.
37. Thomas, Lauren. "Nordstrom Shares Tumble as 2021 Sales Outlook Disappoints, Retailer Sees Bigger Shift to Rack." CNBC, 4 Feb. 2021, www.cnbc.com/2021/02/04/nordstrom-sees-2021-revenue-up-25percent-as-it-puts-focus-on-rack-e-commerce.html.

quotes (here she will likely be averse to using an intermediary such as a broker when she can do all her research online at her convenience); to use a travel agent for her hotel, car rental, and air bookings; to go to a governmental office to research or inquire about a policy and so on.

Our story so far suggests that digital strongly correlates with the online and mobile channels as distinct from the physical brick-and-mortar store. That is true.

Our delineation so far also suggests that whatever this digital of today is, it likely is required to do well—and as the examples suggest, even survive—as a company today. That also is true.

The examples—both good and bad—and the powerful consumer trends cited so far indicate that the case for digital has existential relevance for companies today. In a majority of consumer-oriented sectors, wrong choices related to digital quickly clear the path to bankruptcy. In almost all industry sectors—I cannot think of a single sector where this would not apply—not doing digital well translates to opportunities lost for revenue and margin growth, for building exceptional experiences for and deeper relationships with customers, and for driving accretion in company valuation.

But digital, especially as it applies to the existential argument, is more than just the online or the mobile channel. As we are about to see in the next few chapters, it represents powerful trends that, in combination, represent an industrial revolution of sorts. And as is true of all revolutions, they are a survival play: Only the fittest, or rather, only the ones that have adapted well to the shift, will survive.

In the last two decades, there have been technology innovations and advances that have increased the technology's reach—the number of lives worldwide that technology can touch. These advances have also increased technology's effectiveness—how well it can address a given human need and deliver results. All this has happened while the price of using technology has gone down.

The insatiable appetite of the modern consumer to use technology, along with the number of things that technology can do, have created powerful patterns of demand in the market that all companies have to respond to. Technology is also no longer only inert. It has developed intelligence and an ability to learn from experience. That is, technology now can learn—and train itself—from how it gets used by its consumers. This means that as its usage increases, technology will only get better with time.

What all this has done is broadened the footprint of value that technology can deliver. Increasingly, non-differentiating—the mundane and routine—activities that do not add significant value are getting taken over by machines. In consequence, the workforce employees have more capacity to work on differentiating and value-adding tasks, increasing the number and quality of pathways for innovation.

As a result of all these developments, digital today is helping companies to innovate; explore and commercialize new business models; increase revenues organically; disrupt markets; generate new sources of demand; and serve existing customers exceedingly well, building loyalty and even advocacy for the brand, while reducing costs.

In the next three parts of this book, we will see this more clearly. In the next part—part two—we will take a look at the market and economic demand patterns that are requiring companies to use digital well. Through a review of all the facets that these demand patterns represent, our appreciation for why digital is an existential argument for a modern corporation will be strengthened.

In the succeeding part—part three—where we take a look at the specific technology and organization model innovations that companies are using today as part of their digital footprint to compete in the market, our belief for the existential argument will be strengthened further. Here we will see how entire backend operations can be shifted to a machine platform, at scale, while the company only pays for the platform services that it actually uses—eliminating the need to make heavy upfront capital investments. We will review the tremendous power of Big Data that has changed the game of how companies make decisions. In this portion of the book, we will also see how the new people and process organization models allow companies to keep an active connection with the pulse of the customer and use it routinely to identify the next unmet customer need that can then be tapped into creating superior products and customer experience.

In part four, we will take a look at specific examples of how value gets created. Here we will be able to appreciate how digital creates value in different types of corporate functions such as sales, marketing, customer service, finance, and HR. Following this, we will see the digital value footprint in different industry sectors. We will be able to appreciate how there is not a single function or industry sector today that is not hinged on digital to create value for the modern organization.

By the time the reader has finished this book, hopefully, she will have an evidence-based appreciation for why digital is required for companies to survive today.

Part Two

Demand-side Patterns that Have Contributed to the Definition and Rise of Digital

Demand-side patterns are those trends in the market or technology space that, if a company does not respond well to, can impact the company's profitability, its growth prospects and may even put it on a path to bankruptcy. On the flip side, companies that respond well to these patterns can see new frontiers of revenue streams, business models, and as the examples of Lyft, Uber, and Airbnb have shown, create entirely new markets. These patterns can be seen as the reasons why companies of all sizes today need to respond to digital.

Most of these patterns have evolved in the last two to four decades and are hinged on innovations in technology.

We now look at these patterns one by one.

Pattern One: Compute power at the consumer end

In terms of computing power, a 1964 supercomputer that cost $56M in 2018 dollars could perform 1 million floating operations per second, or FLOPs, a measure of the computational power of a computing machine[1]. In 2018, a miniature smartwatch costing only $250 could perform the same number of operations at about 3 cents[2]. An iPhone6 costing about $800 could perform the same number of operations for about 16 cents[3].

Yes, that is right.

---

1. Olds, Dan. "SUPERCOMPUTER vs Your Computer in Bang-for-Buck Battle." www.theregister.com, 9 Mar. 2012, www.theregister.com/2012/03/08/supercomputing_vs_home_usage.
2. McCarthy, Patrick. "Infographic: The Growth of Computer Processing Power." *RECOIL OFFGRID*, 2 May 2017, www.offgridweb.com/preparation/infographic-the-growth-of-computer-processing-power.
3. Ibid.

A 2018 Sony Smartwatch, or an iPhone 6, in terms of computing power, runs equivalent to a 1964 Supercomputer, at a fraction of the price.

Not only has computing power increased in strength, but it has also become extremely affordable and reduced massively in physical size. That is, modern gadgets are smaller, more powerful than yesteryears, and come at a fraction of the price.

This suggests that modern computing gadgets, such as mobile devices, because they are powerful, small, and affordable, would find mass adoption.

According to Cisco's Annual Internet Report, globally, in 2018, there were 2.4 networked devices (devices connected to internet networks) per capita[4]. This number is expected to increase to 3.6 networked devices per capita by 2023[5]. By 2023, the number of networked devices is expected to be more than three times the global population[6].

The number of mobile subscribers globally is expected to grow as well: from 5.1 billion (66 percent of the world population) in 2018 to 5.7 billion (71 percent of the population) by 2023[7]. Between the period 2018 to 2023, smartphones are estimated to grow at a compounded annual growth rate of 7 percent[8].

Pattern Two: Faster and cheaper networks

An increase in end-compute power is an aspect of technological advancement that represents point-in-space improvements. These

4. "Cisco Annual Internet Report (2018-2023) White Paper." Cisco, 10 Mar. 2020, www.cisco.com/c/en/us/solutions/collateral/executive-perspectives/annual-internet-report/white-paper-c11-741490.html.
5. Ibid.
6. Ibid.
7. Ibid.
8. Ibid.

improvements are localized in that when one is speaking about improvements in the computing power of mobile devices, or wearable watches, PCs, or tablets, these improvements are restricted to the physical space of the device.

Just by themselves, these improvements do not mean much. The most powerful device on the planet is going to have limited value if it cannot interact with other devices.

This is where networks come in. Two powerful devices connected by a slow network that can move information only at slow speeds are going to frustrate the end-user. It will be like having an expensive automobile with no or poor quality roads to drive on.

Fortunately, the internet bandwidth—a measure of internet speed—has improved over the years. By one estimate, the internet connection speed was around 300 bits per second (bps) in the year 1984; this same speed had increased to around 325,000,000 bps (325 Mbps) in 2019[9]. This implies that the internet speed improved by about 50 percent each year[10].

Comparison of network prices over the years per unit, say per bps, are less readily available. But it is a decent argument that network prices have dropped or have become more affordable for use by the masses. Per the U.S. Bureau of Labor Statistics, prices of internet services and electric information providers were about 22% lower in 2020, relative to 1997[11].

If network or internet provisioning prices have become affordable, then that would suggest that the use of the internet has increased.

That is indeed so.

In 2018, 90 percent of the regional population in North America was using the internet[12]. This number is expected to grow to 92 percent

9. Ibid.
10. Ibid.
11. "Inflation Rate for Internet Services between 1997-2021." www.in2013dollars.com, www.in2013dollars.com/Internet-services-and-electronic-information-providers/price-inflation. Accessed 17 June 2021.
12. "Cisco Annual Internet Report (2018-2023) White Paper." Ibid.

by 2023[13]. Globally, 51 percent of the world population was using the internet in 2018. By 2023, 66 percent of the global population—5.3 billion users—is expected to be using the internet[14].

Pattern Three: Free, highly accessible, high-quality information research

Decades ago, if you needed information, you likely would have had to ask someone for it. Libraries, books, journals may also have augmented your quest for an answer.

Now, you just have to frame the question, and type it in a search bar of an engine such as Google, and wait for a few milliseconds to see the results. You can review the results and fine-tune your question to close in on the eventual answer that you are seeking.

Increasingly, searches on Google return an accurate, or at least a set of highly relevant, responses to your question. Quite often, you do not even have to move to the next page of your search results. Chances are that the results shown on just the first page would suffice.

There are a few factors here that are contributing to the increased hit rates on your information searches. The first is the availability of matching content. With physical books getting digitized, for instance, from Google, such that individual words can now be searched for on books' pages, the amount of information that is available—and searchable—to potentially answer your question and whet your appetite has increased by orders of magnitude. Available content has increased not only through higher accessibility of print media but also through the availability of other media types such as videos, and social media messages posted online, most of which are searchable. Thus, for instance, searching for opinions of an esteemed person or a celebrity on a topic

13. Ibid.

14. Ibid.

can quickly inform a curious web user. For example, a person who has just woken up in the US can quickly learn about—and form her opinion on—a cricket team debacle that took place while she was asleep by quickly sifting through search results to her well-formed question.

The second contributing factor is better engineering performance of search engines such as Google and Bing. From an engineering perspective, search engines use three core capabilities: crawl, index, and rank. Crawling entails the content discovery process where the engines send out software programs known as crawlers or spiders to traverse the internet for content that can be anything from webpages, videos, PDF files, images, and so on. The content that the crawlers find is then indexed, meaning that it is organized for display as results to a relevant query. Finally, the content that can be potentially displayed is ranked to indicate the order of most relevance to a search query. With this three-pronged design, search engines today can return relevant results fast.

Increased hit rates have, without a doubt, contributed to the success of search engines. There is another factor that cannot be ignored when analyzing search engine success. It has to do with access. The fact that you can launch and complete your search from your mobile device means that you do not have to wait to reach home and plug in your laptop to do the search. The search is accessible from all your devices, 24x7. This, in addition to increasing the productivity of a customer, employee, or a just-curious researcher, also means that now the search engine has increased its universe of matches that can inform any future searches. That is, the more the search engines get used, the better positioned they become to serve their intended purpose well.

Search engines have come a long way since 1990, when by one account, search engine history started[15]. Then, there was a site called Archie that hosted an index of directory listings that could be downloaded[16]. Those directory listings served as outputs from the earliest

15. "The History of Search Engines—A Timeline." www.wordstream.com, www.wordstream.com/articles/internet-search-engines-history. Accessed 17 June 2021.

16. Ibid.

versions of search engines are a far cry from the nifty search results presented by search engines of today.

It will be good to quantify measures of this search engine success. Direct measures are hard to come by. That said, there are other indicators that can highlight success.

Google, which in 2019 had over 90 percent of the search engine market share worldwide, handled about 3.8 million searches per minute on average across the globe in 2019[17]. About 15 percent of searches that are performed on Google every day are new[18] (this suggests that as these new searches become part of the indexed repository, the ability of the search engine to answer more and more of the questions that come its way, and do it well, increases, further increasing the quality of information search).

Search engines such as Google monetize their search engine capabilities by selling advertisements that are related to search keywords. As a searcher—who is also a consumer—types in her keywords for a search, advertisements relevant to her search are shown to her. These search advertisements are tied to the key terms that brands bid on and that the user types in her search for information[19]. In 2019, brands were expected to spend more than $53 billion in the US alone on digital ad spending related to Search[20].

All this suggests that not only is the quality and speed of information search getting better, it is also leading to an attractive commercial upside for companies whose platforms run the searches.

---

17. Skai. "How Many Google Searches Per Day Are There? | Useful Search Metrics for Marketers." Skai, 16 June 2021, skai.io/monday-morning-metrics-daily-searches-on-google-and-other-google-facts.
18. Ibid.
19. Ibid.
20. Ibid.

Pattern Four: Emergence of new consumer demographics with a high affinity for technology

So far, we have appreciated that devices in the hands of the consumer are computationally powerful and that they have also become highly affordable. We have also learned that the networks that can connect these computational devices have also become more powerful in the sense that they can carry higher amounts of information traffic at cheaper price points.

But all these computationally savvy devices connected by ever so powerful networking infrastructure would amount to little business value if there are no consumers who use them actively.

This is where the powerful new consumer demographics—the Millennials or Gen Y, and the Gen Z—about whom we learned in the last chapter come into play. Recall that the Millennials are those born between the years 1981 and 1996, while the Gen Zers are those born between 1997 and 2015. In the US, Millennials and Gen Z account for 42 percent of the total population.

Globally, in 2020, approximately 1.8 billion people, or 23% of the world population, can be considered as belonging to the Millennial segment[21]. Of this global number, 60% (1.1 billion) reside in Asia; 16% (300 million) in Africa, 8% (150 million) in Europe, and the rest 250 million or so in North America and Oceania[22].

Gen Z comprise 30% of the world population, about 2 billion in number[23].

---

21. "Millennials Demographic Change and the Impact of a Generation." MSCI, www.msci.com/documents/1296102/17292317/ThematicIndex-Millenials-cbr-en.pdf. Accessed 17 June 2021.
22. Ibid.
23. Mccrindle. "Gen Z and Gen Alpha Infographic Update." McCrindle, 27 Apr. 2021, mccrindle.com.au/insights/blogarchive/gen-z-and-gen-alpha-infographic-update.

Together, Millennials and Gen Z comprise about 53% of the world population. Both are technically savvy, and when it comes to key social and policy issues, at least in the US, the two groups think alike[24]. While Millennials are considered the first technology natives, born as they were during the time that the internet developed, Gen Z are considered digital natives since they do not know of a world without smartphones[25, 26].

So what we have here are two technologically and digitally savvy groups—53 percent of the world population—who love to use technology, and what they have in their hands are increasingly powerful devices that can communicate to their *digitoverse* (digitally connected universe) through increasingly expedient and cheaper networks. And, these two demographics that think alike also care about what is happening around them in terms of social, environmental, or political issues, and they do not hesitate to be vocal, quite oftentimes on their devices.

This combination of demand patterns—powerful, well-connected devices in the hands of technology and device-loving new and powerful consumer demographics—the Millennials and the Gen Z—is the single biggest reason why brands today need to respond to digital, and well.

Pattern Five: Ubiquity of data

If you give the digital and technology native generations—the Millennials and the Gen Z—powerful and attractive devices that are well connected, what will flow is data. Lots of it.

---

24. Parker, Kim, and Ruth Igielnik. "On the Cusp of Adulthood and Facing an Uncertain Future: What We Know About Gen Z So Far." Pew Research Center's Social & Demographic Trends Project, 30 May 2020, www.pewresearch.org/social-trends/2020/05/14/on-the-cusp-of-adulthood-and-facing-an-uncertain-future-what-we-know-about-gen-z-so-far-2.
25. "Millennials Demographic Change and the Impact of a Generation." Ibid.
26. Parker, Kim, and Ruth Igielnik, ibid.

Now, just the devices by themselves are likely not going to be of any use. What makes them productive as powerful consumer tools is what is on them. That is where the apps that are on mobile devices come into play. Apps such as Facebook, Twitter, Linked In, WhatsApp, and so on serve as the social media channels that consumers use to research a brand as well as its product and services. The consumers also use these channels to solicit and review opinions that others have of the brand and its products, the brand's reputation in general, and after they have made the purchase, share their own opinion on their usage experience. Apps such as those offered by the e-tailers (companies offering their products and services over the web or mobile) allow for consumers to additionally execute the purchase using a payment gateway.

That is, the modern consumer can execute her path to purchase from soup to nuts over the web or mobile, using the powerful gadgets—the app loaded devices—in her hands.

Consumer interactions with your brand's mobile app or the web portal (the eCommerce site where customers can execute their purchase of your products) can be electronically cataloged. This allows you to track your customers' experience with your brand across all the touchpoints. Customers can upload videos of your product on social media channels. They can share with their friends and their network if the product delivery experience was good or bad. Experience with your brand and your products, as seen from the eyes of your customers, is cataloged over a myriad of channels: social media, your company's websites, third-party websites, and so on.

Chat-friendly consumers with access to multiple channels on powerful devices connected over efficient and inexpensive networks mean that a lot of data gets generated in the process.

It is useful to lay down a scale for understanding the extent of data that gets generated in modern times. A byte represents a fundamental digital footprint of data. It is made up of 8 bits. A bit—a short form for "binary digit"—is the most basic unit of data: 1 or 0[27].

---

27. Thomas, Patrick. "Defining an Exabyte." Backblaze Blog | Cloud Storage & Cloud Backup, 9 June 2021, www.backblaze.com/blog/what-is-an-exabyte.

A kilobyte is one thousand bytes, a megabyte is a million bytes, a gigabyte is a billion bytes, a terabyte is a trillion bytes, a petabyte is one thousand terabytes, an exabyte is a thousand petabytes, and a zettabyte is a thousand Exabytes[28].

For perspective, a kilobyte represents the volume of data in a paragraph of a text document; a megabyte, a small novel; a terabyte, all the X-rays in a large hospital; an exabyte, about one-fifth of all the words people have ever spoken[29].

In 2003, researchers at the University of California estimated that all of the data created till that date and since the dawn of humanity amounted to 5 exabytes, which is equivalent to 37,000 times the total information stored in the Library of Congress[30].

By the end of 2019, the internet had around 4.6 billion users around the world, compared to 3.8 billion in June 2017[31]. In 2020, YouTube received 300 new hours of video each minute and had about 1 exabyte of data on its servers[32]. In 2020, people shared more than 100 terabytes of data on Facebook daily, and every minute users sent 31 million messages[33]. In 2020, every person on the planet was estimated to be creating 1.7 megabytes of information each second[34].

In 2017, there was estimated to be 2.7 zettabytes (ZB) of data globally in the digital world; this number reached 4.4 ZB in 2019 and is expected to reach 44 ZB in just one year[35]. By 2025, this data is predicted to grow to 175 ZB, which is 35,000 times the amount of information created from the dawn of humanity till 2003[36].

---

28. www.eecis.udel.edu/~amer/Table-Kilo-Mega-Giga---YottaBytes.html. Accessed 5 July 2021.
29. Ibid.
30. "The Exabyte Internet." United States Department of Justice, 25 June 2015, www.justice.gov/atr/exabyte-internet.
31. Galov, Nick. "77+ Big Data Stats for the Big Future Ahead | Updated 2021." *HostingTribunal*, 19 Jan. 2021, hostingtribunal.com/blog/big-data-stats/#gref.
32. Ibid.
33. Ibid.
34. Ibid.
35. "The Big Data Facts Update 2020." *NodeGraph*, www.nodegraph.se/big-data-facts.
36. Ibid.

This is the Big Data—data that comes in volumes unknown to man before and at tremendous speed. The reason it is relevant for corporations is that if collected, analyzed, and acted upon in time, it can provide powerful leverage to companies to drive differentiation in how they serve their customers, design and launch their products, and cater to new and existing markets. We must note here that it is not only the customer channels that are the source of this Big Data. This data also originates from machines, and products, that have, for instance, sensors built into them and that can communicate with a corporation's information processing backbone.

As can be expected, most of this Big Data does not get used by corporations. By one estimate from Forrester, between 60% and 73% of all data collected by an enterprise remains unused for further analysis[37]. This represents a tremendous opportunity for companies to not only drive higher differentiation in the market but also to leapfrog the competition. That is where Artificial Intelligence and Big Data Processing can help, and that is what we look at next.

Pattern Six: Breaking through of Artificial Intelligence and Big Data processing

Artificial intelligence or machine learning works something like this:

Say you have a picture of a rose in a digital format. Further, assume that this picture is 100 x 100 pixels (a minute area of the display on the screen). Now, each pixel has a set of three numbers, each ranging from 0 to 255 and signifying how the color red, green, or blue is associated with the pixel. Thus, a set of 25, 150, and 240 associated with a pixel represents that the color red has "an extent" of 25, color green an extent

37. Ibid.

of 150, and the color blue an extent of 240—the three numbers specifying the digital color presentation of the pixel. Specifying these three numbers—the RGB distribution—for each pixel in our 100 x 100-pixel picture provides a digital—and mathematical—representation of our rose. We have used a 100 by 100 matrix for simplification. In reality, the number of pixels used to represent a picture can be much higher.

A set of 30,000 data points or numbers (10,000 pixels, each with a 3 number representation for color) then become a mathematical representation of a picture. If a set of these numbers represents a rose, we give this set additionally a label of 1. If the set, and the corresponding picture, do not represent a rose, we label that set with a value of 0.

This way, we can mathematicize a great number of pictures and deem them to be either representing a rose or not. What we will have in our hands is a labeled data set: on one side, we have 30,000 numbers representing a vector of information for the picture at hand, while on the other side, we have a label of either 0 or 1.

Having created this data set, what we can now do is train a machine-learning algorithm to determine whether a given row of data represents a picture of a rose or not. This training really is a choice of an algorithm, defining how many parameters (the jargon word is features) the algorithm is going to have, and then developing a model that selects the best value for each parameter or feature. This selection minimizes the chance of an error: Not able to detect a picture that is a rose when it is, as well as calling a picture a rose when it isn't. Conceivably, the more data you feed into this model generation, the better may be the model's prediction accuracy. Similarly, the choice of the right algorithm, along with the right set of features, can also help.

The example we used here was a labeled data set. For each row of our data set, we had a label that indicated what the actual result was: whether the row represented the picture of a rose or not. Another domain of data sets that artificial intelligence grapples with is unlabeled data sets. Here the objective is not to find out whether something is A or B, but to find hidden patterns in the data sets. Predicting whether a customer will churn is an example of a labeled data set, and it represents a form of learning known as supervised learning. Identifying coherent customer segments based on a bunch of customer information points is an example of unlabeled data sets, and it represents a form of learning known as unsupervised learning.

Artificial Intelligence programs, or equivalently the machine learning models, feed on massive amounts of data that needs to be stored somewhere to facilitate analysis. As can also be expected, these programs are also computational hogs. For artificial intelligence to become commercially viable, the supporting technology has to be such that it facilitates storage of massive amounts of data—called Big Data—and its processing at viable costs.

One major technological advance has allowed the processing of Big Data to happen at scale, and at low costs.

It is called MapReduce, which is a programming model first popularized by Jeffrey Dean and Sanjay Ghemawat of Google in 2004[38]. What this model does is take a massive data set that could be as large as multi-terabyte worth and break it down into chunks that are each processed in a parallel manner. This parallel data processing takes place on clusters of computers called nodes. One of the breakthroughs is that these nodes can be commodity hardware. What programmers or data analysts do is supply their own code for 'map' and 'reduce' tasks. The map tasks run in parallel on the chunks of the input data set. The output of the map tasks is then sorted and provided as input to the reduce task. Execution of the map-reduce task combination brings the data analysis to its desired state.

Even though running on commodity hardware, the execution of MapReduce can be robust. This is made possible by software programs called trackers that manage and monitor the execution of component jobs and trackers, ensuring that data and execution integrity is maintained throughout.

Hadoop, a popular usage of the MapReduce concept, is used by companies such as Facebook to process logs of user activity and clickstream data (analyzing where a user clicks can help a company gain a

38. Geebu, Ruchitha. "History and Advantages of Hadoop MapReduce Programming." *Mindmajix*, 7 Mar. 2017, mindmajix.com/mapreduce/history-and-advantages-of-hadoop-mapreduce-programming.

better understanding of their preferences, allowing the company to, for instance, personalize content and provide a better experience to its users; this can contribute to profitability)[39].

The combination of Big Data processing and artificial intelligence is creating strong value levers for companies. Companies can process customer data collected in-house as well as augmented from data available from third parties that, for instance, informs about a customer's financial profile, persona, demographics, social, environmental, and political beliefs, and channel preferences. Analyzing this data, companies can predict customer's lifetime value. They can also predict if the customer is likely to churn. Marketing can then use customers' lifetime value scores, and their propensities for churn, to proactively launch intervention campaigns that can reduce customer attrition.

Big Data Analytics is a powerful emerging trend in the market. Companies are using it to drive higher customer retention, reduce costs of customer acquisition by using superior customer segmentation models, and systemically reduce experiential gaps in their customers' journeys with their brands.

According to a recent Accenture study, 79% of enterprise executives believe that companies that do not embrace Big Data will lose their competitive advantage and could face extinction[40]. 83% have pursued Big Data projects to create a competitive edge[41]. According to PwC, 59% of executives believe that the use of Artificial Intelligence would improve their use of Big Data[42].

Per IDC, worldwide revenues for Big Data and business analytics solutions were estimated to be $189 billion in 2019 and are projected

39. Ibid.
40. Columbus, Louis. "10 Charts That Will Change Your Perspective Of Big Data's Growth." *Forbes*, 23 May 2018, www.forbes.com/sites/louiscolumbus/2018/05/23/10-charts-that-will-change-your-perspective-of-big-datas-growth/?sh=44952ac62926.
41. Ibid.
42. Ibid.

to grow to $274 billion by 2022, at a compounded annual growth rate exceeding 13%[43].

Pattern Seven: Environment-aware products

With the processing chips getting miniaturized while also getting more powerful—and cheaper—and with the availability of high-speed internet connectivity everywhere, it was only a matter of time before the processing power entered the physical body of products.

And as with all processing chips, the business value comes from the use of information.

The most natural scenario is when the products can use sensors to "measure" their environment on variables such as temperature, humidity, etc.; use the software on the computing chips installed within them to process the sensor collected information to perform some basic calculations, or even run some algorithms, and then either auto-maneuver themselves, or share the processed information over the internet with a more centralized computing infrastructure, that then informs the device on how to act and proceed further. This analysis power, if small, can be resident on the products or devices themselves. If the need for analysis is bigger, the product or the device can interact over the internet with a centralized infrastructure that has higher computing horsepower. What this does is create an ecosystem of connected devices or products that are now thriving with a computing heartbeat. This heartbeat allows them to show a higher form of animation through, for example, auto-correcting themselves before the emergent

43. "IDC Forecasts Revenues for Big Data and Business Analytics Solutions Will Reach $189.1 Billion This Year with Double-Digit Annual Growth Through 2022." *Business Wire*, 2019, www.businesswire.com/news/home/20190404005662/en/IDC-Forecasts-Revenues-for-Big-Data-and-Business-Analytics-Solutions-Will-Reach-189.1-Billion-This-Year-with-Double-Digit-Annual-Growth-Through-2022.

local issue becomes a bigger problem—a phenomenon known as predictive maintenance.

The connected device or product—a physical object—is a "thing" in the well-known technology paradigm called the Internet of Things or IoT.

IoT can be thought of as combining the abilities to collect information from a device and its ecosystem, analyze the collected information, and use the analysis informed findings to take action within a device.

The value from IoT is not only through predictive maintenance, where you can predict big problems beforehand and remediate them proactively before incurring huge costs from device or machine failure. The value also comes from collecting product or device usage information—how they are being used—and using that intelligence to inform future product design. Thus, products of the future can meet the unmet customer needs better, differentiate themselves more against the competition, and drive top-line growth for a brand. In other words, IoT as a technology can not only help drive down costs—through, for instance, predictive maintenance—but it can also help grow revenues through better-differentiated products.

IoT has opened the flood gates of information that can—and should—be tapped by businesses for growth.

IDC forecasts that by 2025, there will be 41.6 billion connected devices, or "things", generating 79.4 zettabytes of data[44]. That is, by 2025, there will be upwards of 5 such devices for each person on the planet. Dollarwise, the market for IoT technology generated 100 billion dollars in revenue in 2017 and is forecast to generate around $1.6 trillion in revenues by 2025[45].

It stands to reason that while IoT can generate bytes and bytes of information, that information will not create value unless it is tapped for meaningful use. And this tapping requires that there must be addi-

44. "The Growth in Connected IoT Devices Is Expected to Generate 79.4ZB of Data in 2025, According to a New IDC Forecast." IDC, 18 June 2019, www.idc.com.
45. Statista. "Global IoT End-User Spending Worldwide 2017-2025." *Statista*, 22 Jan. 2021, www.statista.com/statistics/976313/global-iot-market-size.

tional technology that can store the ocean of IoT-generated data and process it in good time at viable costs. And this is where the multiplier effect of all the digital technologies kicks in. The Big Data technology allows for storage and processing of the incoming IoT data at low costs. The use of Artificial Intelligence or Machine Learning technology means that there are algorithms that can not only consume the IoT information meaningfully but can also learn and train themselves to become better with time as more and more data becomes available. Conceivably, the sky is the limit. In principle, at least, as more and more data becomes available from the device usage, better and better products and devices can be designed that are not only functionally superior but also come at lower costs.

Pattern Eight: Higher permeability of barriers to entry

The digital technologies—IoT, Big Data, Artificial Intelligence—are sector agnostic: Any industry sector willing to make investments in them can harvest the value that results. And indeed, the value scenarios from the use of these technologies now encompass all sectors.

When a customer enters the store, the retailer can use the proximity of her mobile to an RFID tagged shelf to make her real-time offers or inform her about product outages and make recommendations for available alternatives, improving her experience of visiting the store and driving her long term retention. An insurance company provides a configuration engine on its website or mobile app to a prospect so that he can analyze various combinations of insurance products he is interested in while proceeding forward on the path to purchase. This is of great value to the insurance company because now the insurance company does not have to pay the more expensive salaries and commissions to direct field agents or have indirect brokers eat into its margins. This encourages the use of the electronic direct-to-consumer

(D2C) channel—the so-called eCommerce channel—in the channel mix, which in turn reduces the overall cost of sales since the electronic D2C channel can be less expensive than the direct field-oriented sales force, or the retail outlets.

The health care provider can proactively inform the patient when it is time for the next visit. The provider's web portal can be connected to the patient's wearable device and can send an email when a health parameter such as the heart rate exceeds a threshold. Glucose measuring devices could be in conversation with the patient's gym equipment (both examples of IoT) to make sure that the diabetic patient is exercising regularly. Deviations, especially the critical ones, could be sent to the provider who can close the loop by intervening and making a call to the patient to adjust her medications, preventing an expensive down the line potential visit to an Emergency Room. The patient benefits by staying healthy. The provider system benefits by keeping the costs low by minimizing avoidable patient visits.

The health care metrics over a population demographic—collected from IoT, Big Data processing—can be analyzed over time to detect long-term trends or issues.

Asset-intensive industries such as manufacturing or consumer goods can use device collected information to reduce costs from waste and design more efficient machines. Regulation heavy industries such as the pharmaceutical or banking can make the required compliance documentation available to all their employees over the web and mobile to maximize adherence to required procedures and reduce the chance of expensive penalization. They can even create automated workflows that enforce compliance with required procedures.

In short, digital is democratizing. It levels the playing field across sectors. All and any sector could harvest value from it. Not only that, there can be cross-vertical pollination. The IoT deployment for a consumer goods company can learn from how it was used successfully for a pharmaceutical company.

The democratization aspect of digital is not restricted to only industry sectors. It also covers the user population. A core aspect of digitalization is better technology at lower costs. This means that many more people across the globe can now avail the use of digital technology, regardless of their economic status. We learned earlier how internet and mobile usage across the globe is expected to increase with time. As an example, a farmer in India can now use his mobile device to check on the price of the produce that he is going to get in a market channel before he makes a visit to the channel's physical site location. This farmer may indeed be close to the poverty line. The reason he is able to use his mobile to check on the price is that the mobile, along with a data plan, comes at an affordable price in India. A Jio phone—a mobile phone with calling, texting, camera, browsing, and video watching abilities—is available in India for less than $10 apiece[46]. An unlimited data plan with upload and download speeds of 30Mbps costs about $5.46 per month[47]. (This reinforces the point that digital has become a force to reckon with for businesses because not only does it involve the use of the ever-powerful cutting-edge technology, but it also often needs to respond to higher and more pervasive abilities for humans that come at deeply affordable prices.)

That is, the digital, in addition to leveling the playing field across industry sectors, is also democratizing in access. Low price points fuel increasingly universal access to digital, expanding the reach of technology to historically less empowered sectors of the human population.

The third aspect of democratization involves how digital is used. That is, it entails user experience. At the most basic level, when a user gets introduced to a novel way of doing things, and she likes it, it is human nature that she starts to expect the same novelty in all her experiences. Thus, when a user enjoys being able to shop for and order dresses from the web portal or the mobile app, she starts to expect to be able to complete most of her transactions from the mobile banking app as well, without having the need to visit the local branch. Her retail experience is influencing her banking experience. It also prods her to seek telemedicine and not visit her doctor unless she absolutely has to. The expectations from the retailer, the bank, and now the medical pro-

46. "RS 699 Jio Phone Digital Life Ab Sabhi Ke Liye—Get 4G VoLTE Feature Phone at ₹699.Pdf." Jio, www.jio.com. Accessed 2020.
47. "JIO DATA PLAN." Jio, www.jio.com. Accessed 2021.

vider start to show consistency. Her abandoning of the shopping carts without making a purchase is not very different from how she did not review all her medical results available on the portal—in both cases, she was short of time because the time was evening and she had to rush to the gym. Thus, the respective designers of the web portals for the bank, the retail company, and the hospital can do well to learn from each other. They are designing to the preference of the *same* digital consumer after all. This aspect of similar consumer expectations that is more tied to the channel, and is rather independent of the service-providing industry sector, further contributes to homogeneity across sectors.

The end result of this democratization—reduction of the inequality gap across industry sectors, economic strata, and channels; moving each participant toward having an equal voice—brought on by digital has a powerful consequence. Now, the industry sectors or the verticals are less separated from each other. In business school terms, what this means is that the barriers to entry that allowed a vertical to protect its competitive advantages over a long period of time are becoming more and more permeable. Because it is the same digitally empowered consumer that is driving the demand, the brand experiences are needing to be more consumer-centric and needing to shed more and more of their industry sector specificity.

Also, because the progression of digital technology, be it through IoT, artificial intelligence, Big Data Analytics, mobile device computing, or all of the above, is taking place at breakneck speed, it also means that there is a very high level of disruption, and uncertainty in market demand. It is becoming hard for companies—across all sectors—to predict where their next quantum of demand is going to come from. At the very least, just reliance on the traditional demand-predicting models, which mostly use historical data and patterns, is no longer sufficient to be successful as a company, and indeed is very dangerous. Consumer behaviors are always changing. Consumers demand ever-increasing value at lower and lower prices. How is a brand supposed to respond?

Rita McGrath, a professor at Columbia Business School, suggests from her research that the old model of creating industry-specific competitive advantages and sustaining them over a period of time is losing relevance for defining the modern business strategy[48]. For the simple reason that industries of modern times are no longer stable[49]. Instead of competing in a homogeneous industry sector where all the players are facing the same set of external forces, the competitive landscape for the modern corporation entails the forces—technology and customer experience, for example—that are more or less the same across all verticals[50].

CVS, a leading pharmacy chain in the US, acquired Aetna, an insurance company, in 2018[51]. The connecting dot was superior customer experience and lowered costs[52]. CVS now also offers health check-ups and screenings, traditionally the domain of medical service providers. Amazon, the electronic commerce giant, is also a leading provider of cloud services, a domain of the information technology sector. Amazon is also producing content for home viewing. Amazon also happens to own the Washington Post, a leading national newspaper. That is, the electronic retailer, in addition to retail, also has a footprint in IT, as well as the Media and Entertainment industry sector. Apple, the mobile and personal computer manufacturer, is also producing content. It also has plans to launch a self-driving electric car by 2024[53]. Thus, in the sector of Media and Entertainment, the likes of Apple and Amazon are competing with the likes of Netflix, Sony Entertainment, and Disney.

---

48. McGrath, Rita Gunther, and Alex Gourlay. *The End of Competitive Advantage: How to Keep Your Strategy Moving as Fast as Your Business.* Harvard Business Review Press, 2013.
49. Ibid.
50. Ibid.
51. LaVito, Angelica. "CVS Creates New Health-Care Giant as $69 Billion Merger with Aetna Officially Closes." CNBC, 28 Nov. 2018, www.cnbc.com/2018/11/28/cvs-creates-new-health-care-giant-as-69-billion-aetna-merger-closes.html.
52. Mathews, Anna Wilde, and Aisha Al-Muslim. "CVS Completes $70 Billion Acquisition of Aetna." WSJ, 28 Nov. 2018, www.wsj.com/articles/cvs-completes-70-billion-acquisition-of-aetna-1543423322.
53. "Apple Reportedly Plans to Build a Self-Driving Electric Car by 2024, but It Will Face Stiff Competition as New EVs Flood the Market." *Business Insider*, 22 Dec. 2020, www.businessinsider.com/apple-electric-car-face-stiff-competition-from-growing-ev-space-2020-12?international=true&r=US&IR=T.

As the digital becomes more and more viable as a commercial option for companies, we can expect the democratization it brings to consumers and industry sectors to increase. The resulting convergence in industry sectors will also increase. The digital consumer will occupy the strongest center of gravity in the economic ecosystem. Like a massive star with a strong gravitational pull, the sectors will revolve around the consumer. The most viable strategy will be the one that delivers exceptional customer experiences. And it doesn't stop there. The strategy will need to systemically harvest gaps in customers' experiential journeys with the brand—across all touchpoints—and continually mitigate them by altering channel design and product features, and offering better products and services.

(It is not difficult to imagine or even predict the following consumer utopia:

Years from now, there will be consumer circles where access will be gated (similar to gated communities) and subscription-based. Higher premiums paid by customers will mean beyond exceptional experiences. In this circle, a consumer can shop for apparel; manage her investment, insurance, mortgage, and savings portfolio; manage the entirety of care for her family's medical needs from pharmacy provisioning, to provider engagement, to insurance premium and claim reimbursement; navigate her social and dating options; manage her wardrobe so that she has dress recommendations for the date she plans to have; manage her reading, streaming and content needs; manage her entertainment and travel options and schedule. This premium circle will be elite not because it offers many easy-to-use options to manage multiple things but because of the quality of its predictions. The artificial intelligence models that it uses to predict consumers' behavior will be so specific to the spatial, time, moment, and emotional needs of its individual consumer, so intimately intertwined with its human that it will become the alter persona of the human master it serves. The consuming user of these services will be interacting with this ecosystem not through fingers or voice but via thoughts. Yes, the mere conception of an idea in a consumer's mind will lead to its execution.)

Pattern Nine: Changing patterns for corporate growth

Harvard Business School professor Andrew McAfee and MIT professor Brynjolfsson, in their study published in the Harvard Business Review in 2008, analyzed all publicly traded U.S. companies in all industries over prior decades[54]. What they found was that since the middle of the 1990s, when the internet really took off, and enterprises started to embrace widespread use of technology in their mainstream operations, competition increased its pace. The speed of competition in industries reached unprecedented levels such that there were more "winner-take-all markets". The distance between the leaders and the laggards increased. And this dynamic of the leader leaving the laggard far behind was greatest in those industries that were more IT-intensive. McAfee and Brynjolfsson opine in their article that there are a number of possible reasons for this accelerated pace of competition, such as M&A activity, access to global markets, success from R&D. However, they found that a much more powerful contributing factor to the observed increase in the pace of competition was the increase in IT investments and the effect those investments had on companies' success. With digitization—converting manual processes to technology-automated or technology-enabled processes—companies could replicate their business processes quickly with ease and higher levels of fidelity.

McAfee and Brynjolfsson's study suggests that technology has a higher impact on corporate growth than the traditional growth levers such as mergers and acquisitions, global market reach, and R&D.

The period of 1990 to 2008 can be viewed as the phase when companies after the internet became commercially viable, focused on digitizing their core operations and enabling their organizations to per-

54. "Investing in the IT That Makes a Competitive Difference." *Harvard Business Review*, 1 Aug. 2014, hbr.org/2008/07/investing-in-the-it-that-makes-a-competitive-difference.

form in an internet-connected world. The impact of digitization in this phase covered the shift away from the manual and toward automation in processes such as for finance, sales, marketing, human resource management, product design, and manufacturing.

Around the same time, the next wave of technology arrived.

In 2004, Google published its paper on how it uses MapReduce—a programming model that allows for large-scale computing to take place in parallel and over commodity hardware[55]. MapReduce can be seen as opening the doors for processing Big Data. Prior to MapReduce, companies would need to use a massive supercomputer to run a huge set of complex computing operations. With MapReduce, they could run the same set of operations on a network of inexpensive personal computers, bringing the costs down drastically.

In 2006, Amazon launched Amazon Elastic Compute Cloud, a technology infrastructure service that allows its users to develop IT backends that can scale up or down with business demand, and the company using these services only pays for the infrastructure capacity that it uses[56]. That is, now, the companies do not have to purchase massive and expensive servers, storage, or networking capacity. They just launch the servers they need on the cloud. Not only that, but the management also of these cloud servers is done by the cloud service provider, and in most cases, this server management is machine-automated. What this means for corporations is that they no longer have to allocate capital budgets to buy infrastructure capacity. The use of infrastructure capacity on the cloud now became a variable expense: The quantity of usage determines the price. Further, now, expensive human resources of the company can be deployed to focus on differentiating tasks—the non-differentiating work, for example, of managing servers and the infrastructure-related software that runs on them, could be outsourced to cloud service providers. The cloud service pro-

55. Paul, Ryan. "Google's MapReduce Patent: What Does It Mean for Hadoop?" *Ars Technica*, 20 Jan. 2010, arstechnica.com/information-technology/2010/01/googles-mapreduce-patent-what-does-it-mean-for-hadoop.

56. "Search for: When Did Amazon Launch AWS?" Google Search, www.google.com. Accessed 2021.

viders, because they were good at managing infrastructure, developed economies of scale because they were servicing many customers. The end result was that the cost of managing technology infrastructure came down, shifted from a capital to a variable expense, and the talent in-house could be repurposed to focus on growth. In other words, the digitization of the period from 1994 to 2008 could now be run on steroids. (Amazon's launch of its Elastic Compute Services in 2006—Amazon is one of the leading cloud service providers in the market today—serves as a milestone in the widespread commercialization of cloud; Amazon is considered as the first to market such a service[57].)

The field of artificial intelligence (AI) research was founded at a workshop on the Dartmouth campus in the summer of 1956[58]. Since then, AI saw periods of disillusionment, and funding went down. Known as AI winters, these periods were two: 1974-1980; 1987-1993[59]. Around the first decade of the twenty-first century, breakthroughs started to appear that have sustained progress. These breakthroughs came from the higher quality of AI algorithms. But the fact that these algorithms could now run on numerous parallel computers at a low price also helped. The availability of massively parallel computing at inexpensive price points meant that these algorithms could feed themselves on massive amounts of data. To get better. And better they did get. In 2007, using AI, a team achieved autonomous navigation for 55 miles in an urban setting obeying all traffic laws[60]. Around 2012, scientists at Google deployed machine learning algorithms on a set of inter-connected 16,000 computer processors and let this software-hardware nexus run loose on the internet to learn on its own[61]. After scanning 10 million digital images from YouTube videos, this AI

57. Ibid.
58. "History of Artificial Intelligence." Wikipedia, en.wikipedia.org/wiki/History_of_artificial_intelligence. Accessed 15 Jan. 2021.
59. Ibid.
60. Ibid.
61. Markoff, John. "In a Big Network of Computers, Evidence of Machine Learning." *The New York Times*, 26 June 2012, www.nytimes.com/2012/06/26/technology/in-a-big-network-of-computers-evidence-of-machine-learning.html.

robomon learned to detect a picture of a cat[62]. In other words, the field of AI was starting to show accuracy and commercial viability.

Since then, AI has taken off at a commercial scale. AI is now involved in natural language processing, understanding human speech, facial recognition systems, predicting customer churn, providing higher quality customer segmentation, image searches, and so on.

The point is that since 2008, with the advent of MapReduce, Cloud, Big Data, and AI technologies, the investments in technology are not only going to help widen the gap between the leaders and the laggards in competition, as McAfee and Brynjolfsson found out, they are likely going to be a play (as also pointed out by examples in the 'existential portion' of this book earlier) required for companies to simply survive.

Pattern Ten: Higher Darwinian aspect of external shocks

External shocks have existed for business since the time man executed his first trade. Responding effectively to these shocks can prevent a business's demise and even create new opportunities for growth. Weaknesses in response, as the retail bankruptcies from Covid—a recent external shock—discussed earlier have suggested, can be fatal.

What digital has done in the arena of external shocks and how businesses need to respond to them is that it has shortened the path of response. The time available to businesses to respond to the shocks is reduced. And any gaps in capabilities can get amplified on effects, that is, weaknesses in front of shocks now perhaps carry a heavier price. In other words, the Darwinian process of survival of the fittest in the world of corporations now in the age of digital has become more selective, and it carries a shorter cycle time.

62. Ibid.

There is emerging consensus that the recent external shock, the Covid pandemic, has shortened the path available to businesses to transform digitally. The natural outcomes of digital transformation that were bound to happen anyway now have a need to happen sooner. Human demographics such as the Gen X and the Gen Y with a natural proclivity to use things digital, with powerful mobile devices in their hands, connected at good and relatively inexpensive network speeds, cling more to their smartphones when it has become dangerous to venture outside. Thus, the online and mobile use for executing normal business transactions has risen during the time of the pandemic.

But like all powerful natural forces, digital as a force impacts not only mortality, but also birth.

While it is true that businesses that were digitally unprepared or that had not transformed sufficiently, for instance, in their digital commerce capabilities, or that had not prepared themselves adequately to address the new and modified consumer behavior, saw a rapid demise when the Covid shock hit, it is also true, that businesses who on the other side were digitally prepared, saw revenue increases. We saw earlier examples of companies such as Walmart and Target, who saw favorable outcomes during the time of the pandemic because they were digitally prepared.

The aspect of new opportunities from digital is not restricted only to consumer behavior. It also includes the shifting workplace demographic. Increasingly, more and more workers during the time of the pandemic are working from home. Thus, companies have a need to empower remote-based worker productivity. According to a 2020 report, companies such as Facebook expect as many as half of their employees to be working remotely in the next five to 10 years[63]. The adoption of worker productivity tools has picked up pace during the time of the pandemic. Zoom video calling is an example. Another example is Slack, a workforce collaboration software. Around 2016, the software company had 4 million users; by 2020, this number increased

63. "Facebook Expects Half Its Employees To Work Remotely Permanently." NPR, 21 May 2020, https://www.npr.org/sections/coronavirus-live-updates/2020/05/21/860382831/facebook-expects-half-its-employees-to-work-remotely-forever.

to 12 million[64]. Towards the end of 2020, Salesforce, the CRM giant, acquired Slack for $27.7 billion[65]. This acquisition is one example of how accounting for the changed workforce behavior as created or accelerated by the pandemic is part of companies' future vision for growth.

It is likely that the winners from the pandemic external shock will be companies that, having survived the scare, will strengthen their digital capabilities that allow them to serve their customers better, continually identify their next sources of demand, develop differentiated products and customer experiences, while attracting and retaining a digital workforce talent pool that increasingly is in short supply.

But what specific capabilities do a company need to build out digital? That is what we look at next.

64. Newton, Casey. "How Microsoft Crushed Slack." *Platformer*, 3 Dec. 2020, www.platformer.news/p/how-microsoft-crushed-slack.
65. Griffith, Erin, and Lauren Hirsch. "Salesforce to Acquire Slack for $27.7 Billion." *The New York Times*, 4 Jan. 2021, www.nytimes.com/2020/12/01/technology/salesforce-slack-deal.html.

Part Three

Levers that Allow Companies to Respond to the Demand Patterns for Digital

The demand patterns that we looked at in the last part are the reasons why companies today need to respond to digital. They are the market, technology, and consumer trends that just cannot be ignored. These patterns represent the opportunities as well as threats that the companies need to address in order to survive in today's hyper-connected heavily-consumerized global economy.

Now, we look at the specific levers that companies can use to respond to these demand patterns. In doing so, we will have created a holistic framework of capabilities that comprise digital, and that can position a company to tap into the market opportunities created by digital.

This part of the book directly answers the question: What exactly is digital?

Broadly, the required capabilities that comprise the footprint of digital use three building blocks: Technology; Organization models for technology and software design; and Organization models for people, process, and management.

We take a closer look at each of these building blocks next.

Building Block One: Core Components of Technology

Reflecting the advances of mostly the last two decades, the technology components that make up a commercial deployment of digital today cover everything from automation, Cloud, Big Data processing, analytics that goes beyond the traditional business intelligence of the '90s, IoT, and of course artificial intelligence and machine learning.

In combination, these components can help a company develop scalable backend infrastructures that show elasticity with the end-user demand—when the user demand goes up or down, the infrastructure capacity provisioning is automatically increased or decreased—saving precious investment dollars. The technology components can also enable a company to be in a constant listening mode to the sentiments of its customers as well as its perception in the marketplace and tap into the gathered intelligence to make timely strategic and operational decisions. Further, and this perhaps is the biggest marvel, most, if not all, of these technological advances can be used without breaking the bank. For instance, while the use of the Cloud—that allows for the scalable infrastructure—leverages a pay-as-you-use model—a company pays only for the services and infrastructure capacity it actually uses—the deployment of data gathering and associated intelligence can be done in small increments, reducing the cost of failure.

1.1 Virtualization and Cloud

At the end of the day, all technology runs on hardware. All information is stored somewhere; it is then accessed, retrieved, and processed into a consumable form. The consumption typically informs a decision of some sort.

The hardware that underpins all information and digital technology comprises processors, memory, and storage. These three components can be seen as the core capabilities offered by a single physical computer. Computer storage stores information as bytes of data. Computer processing helps execute operations on the data so that the information inherent in the data can be consumed. Computer memory aids processing: it allows the computer to store the data close to the processing unit so that it does not have to make the long trip to the "storage warehouse" each time.

With the advances in technology, computing processors have become faster, and more memory and storage can be purchased at lower prices. That is, the computer hardware now is not only more powerful than before but also cheaper. But the benefit from powerful hardware gets compromised if only one software application hogs all of it.

The software also has to advance to overcome this limitation. It must be able to run in such a way that one software application can share with other software applications the underlying hardware resources.

Enter virtualization.

Virtualization is an advance in software that creates a layer of abstraction over the underlying hardware resources of a single computer or server: its storage, processors, and memory. What it does is create virtual allocations of hardware resources to individual units known as virtual machines or VMs. Each VM is a virtual representation of the underlying physical hardware resources. The software applications run on the VMs. The benefit of this is obvious and powerful. Now, one can run thousands and thousands of software applications (multiple tenants) on the same set of physical hardware (same rented house). This means you can do more with less.

The type of software that makes virtualization possible is known as a hypervisor. In essence, a hypervisor is a layer of software that coordinates all the virtual machines so that the VMs do not interfere with each other's memory or other resources and play nice with each other. The software applications—the business functionality providing code, the stuff that you are really after—then interact with the VMs to access the hardware resources it needs.

Virtualization is a big deal. It is the core technology that has made the Cloud possible.

And the implications are far-reaching.

First, with the advent of virtualization, an organization gets more value from hardware. On the same set of hardware, it can do more things: it can run more software applications.

Second, if the virtual machines can be launched on-demand, and shut down automatically as well to release the hardware resources, then the server capacity can scale up or down automatically as business user demand varies. In the old days, to account for peak usage volumes, businesses had to purchase required hardware or server capacity upfront. No more. Now, through the cloud technology that uses virtualization, the hardware or server capacity can be scaled up or down automatically as the business usage goes up or down. The virtual machines get launched automatically when the demand goes up, and they are shut down automatically as well when the usage demand subsides. That is, the cloud technology now allows for elastic scaling of backend hardware or server capacity (the scaling is elastic because, like an elastic rubber band, when the demand does down, the virtual machines are shut down as well, and the server capacity is released back to the pool).

Third, the businesses now need only pay for the server capacity they actually use. For instance, around a Black Friday, an eCommerce retailer may end up using a lot of virtual server capacity to accommodate thousands of shoppers that are browsing and shopping on its website. Some weeks after, when the user traffic subsides, the peak server usage can be released. The retailer who is running her eCommerce website on the Cloud pays for 40 instances of giant servers on Black Friday but for only five a week after.

This 'pay for only what you use' aspect of cloud computing has significant implications for corporate planning. Rather than budgeting for and purchasing hardware capacity upfront, now, corporate planners need only pay for capacity as they need it and use it. This clearly leads to operational efficiency since incidences of capacity lying unused decrease drastically or are eliminated completely. But it also has another benefit. Now, infrastructure costs shift from being a capital expense to an operational one, from being a fixed cost that depreciates over time to a variable one. This frees up precious capital for other uses.

The fourth advantage, brought to businesses by virtualization and Cloud, has to do with the good old economies of scale. Businesses care about software applications: The bundles of code that encapsulate business functionality and that are the direct holders of business value. But these software applications need hardware to run on. In the pre-cloud days, businesses had to purchase hardware capacity upfront. Now, through the intermediary of virtualization, the boundary between the business use of software applications and the backend server infrastructure has been clearly drawn. The businesses need only worry about and focus their talent on business issues. The backend gets outsourced to cloud service providers. These cloud service providers purchase massive infrastructure capacity and deploy this capacity throughout the globe. Through virtual machines (multiplier effect), they now can run software applications many times over. These cloud service providers become better and better at what they do, developing economies of scale in the establishment and operation of backend infrastructure.

Many terms are afloat that represent this trend. Infrastructure as a Service, or IaaS, is one that represents the outsourcing of basic infrastructure services such as compute, storage, memory, and networks from a business to a cloud service provider. The cloud service provider charges the business for the infrastructure capacity they use and scales the capacity elastically with the business's user demand. Other terms, such as Software as a Service, or SaaS, offer the entire software functionality such as Customer Relationship Management, or eMail, over Cloud.

Cloud service providers are doing brisk business, representing the simple fact that the advantages of doing Cloud have caught on with all businesses, big and small. The big three are: Amazon Web Services or AWS cloud platform, from Amazon; Azure, from Microsoft; and Google Cloud Platform from Google.

For illustration, Amazon Web Services, the cloud computing division of Amazon, posted $45.3 billion in revenues for the year 2020, a growth of nearly 30% year-over-year[1]. The division posted $13.5 billion

1. Google Search for "Aws Annual Revenue 2020." www.google.com. Accessed 2021.

in annual operating profits, a full 63% of the entire company's operating profits for the year[2].

1.2 MapReduce and Big Data processing

Imagine the following capability: A business is able to continuously take a massive volume of data, break it into chunks, and process each individual chunk in parallel. Further, this processing can be done on commodity machines, avoiding the need to purchase expensive computers. Of course, the processing must be customizable. That is, different businesses for their different needs should be able to provide context-driven processing with ease.

This utopia ofbeing able to process large data volumes—cheaply and efficiently—has arrived through an innovation known as MapReduce.

In 2004, Google published its paper on MapReduce. In essence, MapReduce is a programming model for processing large data sets[3]. A typical deployment allows for context-specific customization on two fronts: one can specify one's own map function that processes a key and value pair to generate a set of interim key-value pairs, and then a reduce function that merges all the interim values tied to the same interim key[4]. To use the mathematical calculus analogy, the first function that the user supplies—the map function—perhaps resonates with the calculus operation of differentiation, while the second one, also user-supplied (the reducer), resonates with that of integration. From another angle, the first function breaks down (analyzes) the input data to create more chaos; the second one synthesizes to produce order and coherence.

Software programs written following the MapReduce paradigm are automatically parallelized and executed on a large cluster of inexpensive commodity machines, allowing the programmers that do not have any experience with parallel and distributed systems to enjoy the ben-

2. Ibid.
3. Dean, Jeffrey, and Sanjay Ghemawat. "MapReduce: Simplified Data Processing on Large Clusters." Google Inc., 2004, www.google.com.
4. Ibid.

efits and resources of a large distributed system[5]. That is, large sets of data now can be processed expeditiously with ease.

Complementing the MapReduce programming model is another technology innovation called the Distributed File System (DFS) that allows for the storage of large chunks of data on commodity machines that can communicate with each other using high bandwidth connections. Thus, while DFS allows for inexpensive storage of—and access to—data while it is getting processed, MapReduce as a programming model allows for software code to be applied to the data. Together, DFS and MapReduce allow companies to tame the big volumes of data that reaches their shores and harvest it to make timely decisions on things ranging from supply chain optimizations; to increasing customer shopping cart conversions; to identifying best merchandise combinations to place on store shelves; to scouring research and development data for the next most promising drug molecule to explore and so on.

Another remarkable thing about MapReduce and Distributed File System innovations has been that their commercialization uses open-source—freely available software. This means that these innovations continue to benefit from the creativity of software developers who contribute their talents mostly pro bono. Commercial models of these innovations now use the Apache Hadoop technology ecosystem that is open-source software. It is interesting to note that Apache Hadoop's software components leverage Google's original papers on MapReduce and Distributed File System[6].

1.3 Internet of Things Fueled Analytics, Automation, and Control

Any device that serves a consumer or an industrial purpose such as a wearable watch, a thermostat at home, a machine deployed in a manufacturing operation, or a commercial delivery truck, can be embed-

5. Ibid.
6. Apache Hadoop. en.wikipedia.org/wiki/Apache_Hadoop.

ded with a sensor that monitors the device's environment, and some software that can process the information collected by the sensor. This sensor and software combination in the device can be supplemented with a bi-directional network that communicates with—and receives instructions from—a powerful backend computing and analytical infrastructure. This information supply chain, when enabled, can lead to powerful things. For one, now, the device can self-control itself by receiving correcting instructions from an automated backend intelligent infrastructure. Thus, expensive errors and machine failures can be avoided before they become huge costs. Also, the machine or device usage can be monitored to design better machines—with a lower cost of operations or more user-benefiting features—in the future.

The device that is embedded with the sensor and the software is known as, well, the "thing", and the information-connected network that this thing supports is known as the Internet of Things, or IoT.

The device just with a sensor and software is of no use. What makes the business value come out of this "intelligent" device is its connection with a backend infrastructure that can analyze the information connected by this device for some value. Most often, this value comes through predictive maintenance (device auto-corrects either itself or some process step before it becomes an issue), through better product design, but also through being able to serve customers better, e.g., by meeting their unmet needs at lower or zero additional price.

The backend infrastructure needs to be able to process massive volumes of information. To see this, just consider that a sensor from just one device may be streaming data continuously throughout the day. The volume of data that needs to be harvested for insights then multiplies manifold when hundreds or thousands of devices are communicating. Further, time-oriented analysis, such as for insights over usage that spans months, quarters, or years requires the processing of even more data. In short, for the Internet of Things to produce value, information processing has to be able to operate at scale, be robust (not fail on volume), and do so inexpensively.

This is where Cloud comes in handy for IoT. Through Cloud, the massive volumes of device data can be processed inexpensively at speed.

Cloud Service Providers such as Amazon Web Services provide a capability called Edge Computing. Here, there is a cloud-provided analytics-processing infrastructure—the Edge platform—that is in close physical proximity to the device-sensor ecosystem. This Edge infrastructure can process information that requires short decision-making turnaround times. This Edge infrastructure then communicates with a centralized computing infrastructure—more distant, but also on Cloud—that has the massive horsepower to perform more heavy-weight computations such as for detecting patterns over time.

IoT can be segmented into a few categories: Consumer, Commercial and Industrial.

Consumer IoT covers examples of wearable devices that are used to monitor a customer's health from her fitness profile. These wearable devices are then used to communicate with a healthcare provider for preventive management of diseases such as diabetes. Down the line, these wearable devices can be used to—carefully—market product and service offerings to the consumer.

Commercial IoT focuses on business process efficiency. For instance, clock-in and clock-out frequency patterns from employees can be used to make decisions about air-conditioning usage in an office or provide incentives to employees to prevent attrition. Management of a fleet of commercial trucks, or car rentals, can use insights from telemetry data provided by the vehicles to improve utilization and on-time availability.

Industrial IoT (IIoT) involves machine-to-machine communications in industrial settings. Here, the sensors provided in machines are used to provide an additional layer of automation and control. This is being termed as the fourth wave of the industrial revolution. Smart power grids and smart cities are examples. RFID-enabled shipping cargo—another example of IIoT—allows for effectively managing a supply chain, reducing waste, and providing visibility of expected merchandise to downstream operations. Later in this book, when we review the value created from digital, we will see specifically how IIoT is used to create value in the Manufacturing industry sector.

1.4 Artificial Intelligence and Machine Learning

Mathematics enabled the subject of Physics to model the observed world. Through mathematical tools of algebra and calculus, the rate of change in the worldly phenomenon could be modeled. Thus, fields such as fluid dynamics (modeling of fluids), thermodynamics (modeling of energy), chemical engineering (manipulating conversion of chemicals), and so on emerged.

The field of Artificial Intelligence (AI) is no different. It uses mathematics heavily to represent the environment it learns from, its learning model, what it learns, and its predictions. In essence, AI develops algorithms that are fine-tuned to provide better predictions.

Machine Learning (ML) grew out of AI. It is helpful to learn a good definition of ML. According to Tom Mitchell, an AI scientist: "A computer program is said to learn from experience E with respect to some class of tasks T and performance measure P, if its performance at tasks in T, as measured by P, improves with experience E."[7].

So, for instance, an ML algorithm that is playing chess has an experience E of playing many games of chess, a task T of playing chess, and a performance measure P as the probability that it will win the next game of chess.

The past is learned from to predict the future. The ML algorithm that does it well is deemed to be a well-trained ML model.

All ML feeds on data. The data sets—numerical representations of some observed phenomenon such as features of a customer profile, market segment, and so on—may have either good results or no results. In case the data points in a data set have values for good results, the data set is said to be labeled, and the learning or training—modeling of the best ML algorithm really—that takes place is known as Supervised Learning. The ML model training that takes place on a data set that has no labels—no good result values—is called Unsupervised Learning.

Let us take some examples.

7. "Machine Learning." Coursera's Online Course for Machine Learning, uploaded by Andrew Ng. Accessed Sept. 2019. www.coursera.org/learn/machine-learning.

Given a health profile of a patient, such as information about the patient's age, gender, ethnic profile, and her body mass index, an ML model can be developed that predicts whether a new patient has a good likelihood of having diabetes. This is a case of supervised learning because the model uses existing data on patient profiles mapped to whether patients have diabetes or not (that is, the data the model uses has labels). So, for instance, one can profile the population in the city of San Francisco, capturing a person's health parameters and whether the person has diabetes or not (that is, providing the data its labels). This captured labeled data will then feed into an ML model that can then predict the probability of whether the next person will have diabetes.

Contrast this with a customer segmentation problem. Here a typical data element will have customer demographic and other information such as past transaction history, credit profile, etc., but there is no right result attached to this data element. That is, the data is unlabeled (a case of unsupervised learning). In this case, an ML algorithm will look for patterns in the data. These patterns will be customer segments that represent, for example, similar buyer profiles that then can be specifically targeted.

In the case of supervised learning, where we already know what a correct output looks like, the prediction of the model can be a continuous value or a discrete value. The case of predicting the lifetime value of a customer from a bunch of customer data is an example of predicting a continuous value. In comparison, the case of predicting whether a customer will churn or not is a case of predicting a discrete value. Continuous value predicting algorithms use models known as the regression models, while the discrete value predicting models are known as the classifier models. Both are examples of supervised learning. Note how using the same or comparable set of data, one can use a regression model to predict the customer's lifetime value or a classifier model to predict customer attrition.

The learning of an ML model starts with a hypothesis. This hypothesis is formally a mathematical function that has a bunch of parameters tied to the individual elements of a data point. So, say, a simple model uses a tumor's weight, color, and size to predict whether it is malignant or not (a supervised learning classification problem). The model will have a judiciously selected mathematical function that will have four parameters: one for each of the data's elements or features (a tumor's weight, color, and size), and another parameter that is independent of any data feature[8]. This four-parameter function now becomes the hypothesis for the model that we want to train on the input data about a bunch of tumors. The model is trained on the data by capturing and reducing its error rate. This error rate, known as the cost function, is the difference between the model predicted result and the actual result[9]. For instance, on a known case of a malignant tumor, a comparison is made with the model prediction (that is, whether the model also predicted the tumor to be malignant or not). By systemically measuring this error rate, the model is trained to have the optimal values for its parameters that minimize the error rate or the cost function. Once the cost function has been minimized to the fullest extent possible, the resulting mathematical function is deemed to have the converged or the best set of parameter values. This resulting function can now be applied to real-world data to make predictions.

To increase the accuracy of the model, one can choose a better algorithm, get more data to train the model on, or add more features (columns) to the data set. Quite often, the selection of a better algorithm does not move the needle by much. The choice becomes whether one gets more data (which can be expensive to procure) or add more features that can make the model more complex. There is no one right answer. The emerging consensus is that if the model's performance is such that its predictions are overfitting the data (a case of high variance), then getting more data to train the model on might help[10]. On the other hand, if the model's predictions are under-fitting data (model is showing a high bias), then getting more training data is unlikely to help, and adding more features to the data might be more effective[11].

8. Ibid.
9. Ibid.
10. Ibid.
11. Ibid.

Machine learning models are being used to perform data mining where extensive volumes of data are scoured by the algorithm to generate valuable insights. As an example, mining the web clickstream data (a detailed log of visited web pages, time spent on each page, previous destination, and the next destination) can identify opportunities for producing a better customer experience with the website or for identifying incidents of alarming access.

ML is also being used for recognizing written alphabets, natural language processing, providing auto-piloting for helicopters and cars, and for providing product and movie recommendations on sites such as Amazon and Netflix.

If IoT and social media platforms generate data that companies can use for insights on customers, products, and brands, and if Cloud provides a simple, scalable, and elastic platform to store and process this data, then ML provides the answer to the question: "What to do with this Data?" The ML models are the last mile to value.

1.5 Robotic Process Automation

Through an act called screen scraping, a user or software extracts data from what is displayed on a computer screen. The data present in the specific fields of a web form can be noted down by the human or the software. This collected data can be captured as raw text and analyzed by the software according to specified business rules; the software then invokes other applications for downstream processing. The responses from the downstream systems can be further analyzed by the software to communicate to other systems or humans. This way, the software has been able to read what is displayed on a user's screen, understand it, and use that understanding to invoke a bunch of downstream systems in an orchestrated manner to execute a complete or portions of a business process workflow.

The software that does this is the Robotic Process Automation (RPA). One notices that RPA, in essence, mimics human actions, the actions of an end-user to be specific. In that sense, and for its core feature, it is the monkey of digital technology.

Once the internet became a commercially viable tool in the 90s, and when software programming languages provided the ease of development at scale, companies started to digitize their manual, paper-based operations in a big way. A related maneuver that caught on was that of labor arbitrage: companies began to outsource their business processes to service providers that could run the companies' operations at lowest costs. Per Deloitte's Global Outsourcing Survey, in 2016, 75% of profiled organizations reported that they had achieved their cost-saving targets using labor arbitrage[12].

Using RPA to further increase cost savings by adding automation in the execution of business processes is now the next wave to increase operational efficiency.

But perhaps RPA sounds bigger than it actually is. The grandiose perception perhaps comes from the appearance of the words 'robots' and 'automation' in one name. At the end of the day, RPA is just software. It does not represent the auto-pilot machines called robots that one finds in sci-fi movies or in a car manufacturing plant. And at best—and this is important—it can only mimic what a human or a system user can do. And here also, RPA covers only a subset of human actions that involve interaction with a computer system to enter and process data. As we have noted, RPA can also help application to application interactions where humans are not involved.

RPA's use, understandably, varies across industries because business processes do.

RPA is not best suited for all business processes. Its use excels in areas where the business processes show a few characteristics: They are repeatable, stable, structured, and are not too prone to error. In such scenarios, software modeling can take place to put the human burden of the process on RPA. Here, RPA can read text, launch portions of workflows, effectively reducing the number of tasks that humans have to do.

---

12. Dilmegani, Cem. "What Is Robotic Process Automation(RPA)? Complete Guide [2021]." https://research.aimultiple.com/rpa/, 2021, research.aimultiple.com/rpa.

It is worthwhile to note that business process management (BPM)—the automated orchestration of business workflows—typically encompasses RPA applications. BPM continues to be the master orchestrator that maps closely with a company's business process. RPA then complements BPM by automating human-led tasks of the workflow or backend system-to-system interactions. In this way, RPA further reduces any process inefficiency.

Artificial Intelligence is increasing the number of things that RPA can do. Recent advances in AI have increased the accuracy of optical character recognition or image recognition technologies. Both of these increase the accuracy of screen scraping in turn, allowing RPA programs (called bots) to do more things with higher accuracy.

Most back-office operations are candidates for the use of RPA. In Insurance, for instance, information required for filing claims can be fairly routine, and the steps involved in processing a claim tie naturally to specific business rules. This use case then becomes a good candidate for modeling through RPA. Hand-written forms can be automatically detected by an RPA software program that can spawn required backend workflows. In this scenario, a portion of work that used to be performed by humans, now through the use of RPA, has been taken over by a machine.

Management of account receivables, inventory, order fulfillment, underwriting, procurement, customer interactions in a call center (through chatbots) are other examples where the use of RPA is becoming common.

As AI's success increases, there can be more and more intelligence embedded in an RPA program, and the act of mimicking by software will be less and less of the only imitation. Further, as the adoption of the Cloud increases, the RPA capability can also be deployed on Cloud, and the concept of RPA-as-a-service appears on the commercial horizon.

State of the art in technology is such that while the ability to imitate human actions through software—as RPA does—is recognized, there is little optimism on when, or if, the software, in addition to just replicating human actions, can also perform other crucial human activities: Judgment, and proactive decision making.

It is interesting to note that RPA can offload from humans the front-end activities such as entering the data on a web form. RPA can also invoke backend systems and process their responses. This creates the possibility that an RPA deployment can be end-to-end: it automates and integrates front-office and back-office operations.

The global RPA market was estimated to be $1.4 billion in 2019 and is expected to reach $11 billion in 2027, a growth rate of 34%[13].

The use of RPA in a modern corporation is becoming table stakes. RPA is used to automate mundane backend and front-end processes, driving down the costs of operations—if RPA is done right—over the long term. But there is another benefit. When the mundane tasks are taken over by machines, it frees up a company's employees to focus on more motivating and creative tasks, improving the crucial workforce morale and productivity. It also means that talent can expect that those portions of their jobs today that are low on the differentiating side are likely to be taken over by the machine in the future. Continuous re-skilling becomes the norm.

1.6 Data Streaming and Real-Time Processing

Computer systems interact with only two things: other systems or humans. Each time a system performs interaction with another system or human, an event known as a transaction, data gets generated that represents information of potential value.

Systems such as websites, or customer relationship management systems, of enterprise resource planning systems, are known as online transaction processing systems or OLTPs for short. These OLTPs are sources of data.

This source data carries some value, but it is of limited use. Sure, it is important to capture how much a customer paid for her purchase or what items she bought. Indeed, for public companies and for any company that cares to pay taxes, it is important to capture such informa-

13. Ibid.

tion to, for instance, figure out a revenue basis. A far more important value for an organization comes from insights that the captured data is latent with.

Such powerful insights that businesses can use to, for instance, make a strategic product or sales and marketing decisions, typically require an ability to process data that spans over a period of time such as a quarter, or a year, or many years. Further, these insights require analysis over a much bigger volume of data. And the data needs to cover many dimensions such as customer, products, and so on. Thus, a business may be interested in knowing which products provide the highest margin and for which customers, which customers are the most lucrative in that they buy the biggest volume of products or buy products with the highest margins, which customers are loss-making because they buy less relative to the amount of time they spend over the phone with the customer service representative adding to call center transaction costs. All these questions can be answered only after the captured data is analyzed using a systematic approach.

But of course, detecting such patterns for insights takes time. The data from the transaction systems has to be poured into a reservoir that is conducive to analysis. In technical terms, this means that the data from the transaction source systems has to be extracted, transformed, and loaded into a data warehouse that is designed for deep analysis. Doing all this is the domain of Business Intelligence or descriptive analytics. This portion of analytics generates insights from structured data such as what is captured for a company's orders, customers, inventory, employees, and so on. The insights that descriptive analytics generates typically support operational and tactical decision making, such as is required to support sales forecasts, product decisions, and financial reports shared with the regulators and the general public (if the company is public).

But in modern times, companies also have to deal with unstructured data. This data enters an organization's boundaries from a variety of sources such as social media feeds, IoT devices, training systems, and so on. This data is unstructured because, in a sense, you do not know beforehand the type of data you are getting, and yet you still have to be able to absorb it.

In addition to having to deal with massive volumes of incoming unstructured data, modern organizations are posed with another challenge. Between the event that generated the data and the time the organization has to respond to the event, the gap is exceedingly short, sometimes a matter of only milliseconds.

Consider an example. These days when a web page is loading, there is software that gets activated to generate online bidding for ads. While the web user is loading her page, bids are made for ads in 100-120 milliseconds or even 50 milliseconds[14]. In this short time span, the bidding software system requests bids from all the buyers for an ad spot, deploys criteria to select the winning bid, displays the bid, and collects information related to what happened after the ad was displayed[15].

To achieve such fast response times—a phenomenon known as real-time processing—a few things have to be in place.

To understand this, consider a simple analogy. When you are cooking in your kitchen, how convenient is it when the items you use frequently are placed within reach? Such proximity impacts how much time you take to make the dish, your response time. If these same items were instead placed in a distant pantry, requiring you to make a trip each time you needed them, your dish-making time would increase. The gap between the two response times can increase significantly when we are talking scale: When you have to make many hundreds or thousands of dishes.

In computer processing, the data that comes a system's way and that needs to be analyzed needs to be stored somewhere. There are two options for storage: a disk where it can be stored on a more permanent basis but whose access requires time, or a more proximate location that is closer to the computer processing unit (CPU). Known as the Random Access Memory (RAM), or just memory, this proximate location can drastically reduce processing response times. But it used to be that it was expensive.

14. AWS Amazon. aws.amazon.com/nosql/in-memory. Accessed 23 Feb. 2021.
15. Ibid.

In recent times, RAMs have become cheap. In addition, the CPUs have become more powerful and also cheaper. What this has led to is a technological advance known as In-Memory databases.

As their name suggests, In-Memory databases use the RAM to store incoming data, making the response times to the tune of hundreds of milliseconds possible.

The In-Memory databases have helped to reduce data processing response times. But they still do not fully address the problem of being able to handle massive volumes of data such as is coming from thousands and thousands of IoT device sensors or social media interactions from thousands and thousands of global users. The problem of volume gets addressed through the use of message queues that use the publish-subscribe (known as pub-sub for short) model of information exchange.

To see how the pub-sub model helps with volume, consider another analogy.

A reader typically subscribes to a newspaper or a digital feed for a topic of interest. When information about a topic becomes available from the content provider, it is put on a location; subsequently, the reader, when he is ready and, per his convenience, pulls the information. This so that the content provider is not waiting for the reader to pull the information. And the reader is not waiting on the content provider for when it is ready with the information. The two—the publisher and subscriber of information—can do other things while the information is getting generated or consumed. This model lends itself to scale.

To see the other side, consider the scenario if the reader had had to wait on the content provider. Or if the content provider had to wait for the reader to consume the information before it could move on to generate information on the next topic. In such a scenario, known as synchronous processing, waiting would clearly have become the bottleneck.

The ability of modern organizations to process data from thousands of IoT devices, or information feeds, has become possible through the use of a technology advance known as the Message Queue. The organization can subscribe to the events it finds of interest and that are published on the Queue from the data generating or event producing sources. Once the events of interest are published, the organization can process the data inherent in the events, and if it uses In-Memory databases, the organization can do this processing really fast.

Cloud Service Providers such as Microsoft Azure provide Event Hubs that can ingest millions of event messages per second[16]. They also provide IoT hubs that can handle messages from millions of simultaneously connected devices, providing to and from communications with the device[17].

The combination of event or IoT hubs that can ingest millions of event data units per second and process them in milliseconds using in-memory databases is powerful. It means that a modern organization can tap information from an increasingly interconnected digital universe and generate meaningful actions or insights continuously and in close to real-time. Such is the power of real-time processing.

Trulia, the online real estate site, uses location data to provide information to its users. A change in a city boundary, or county line, needs to be reflected in all analytics services delivered to its customers and managers[18]. Trulia uses real-time processing technology to do this.

A financial services company can actively monitor the streaming user activity data, and use real-time processing to quickly detect fraudulent activity, and take preventive action such as alerting the user. A delay of even a few seconds can be catastrophic in this case.

Wearable devices can stream user activity data to a real-time processing platform that can alert users to a potential cardio event allowing them to take remedial action in time[19].

Ecommerce web giants such as Alibaba use real-time processing to refresh price and availability data from changing product catalogs, and to provide a personalized experience to their users in general[20].

---

16. Microsoft Azure. docs.microsoft.com/en-us/azure/architecture/data-guide/big-data/real-time-processing. Accessed 23 Feb. 2021.
17. Ibid.
18. Lawton, George. "7 Enterprise Use Cases for Real-Time Streaming Analytics." Searchbusinessanalytics.Techtarget.Com, 28 Mar. 2019, searchbusinessanalytics.techtarget.com/feature/7-enterprise-use-cases-for-real-time-streaming-analytics.
19. Ibid.
20. Ibid.

1.7 Structured versus Non-Structured Data

In the case of real-time data streaming, we learned how the proximity of the computer processing unit to data—the choice of in-memory (closer) versus data on disk (farther)—impacts how quickly data processing completes and thus the end-user response time.

Another factor that impacts data processing is how the data is stored. Specifically, whether a structure is used to capture data or not.

Providing a pre-defined structure to data means that you can quickly search an entire database for specific values using a variety of search combinations[21]. This is the biggest advantage that the structured databases—also known as the relational databases such as the Oracle database or Microsoft's MySQL database—offer. You can store data in multiple tables, retrieve data from these tables, and then combine the retrieved data to satisfy your analysis needs. Thus, you can efficiently find out which customers have purchased what products over the last two quarters. Relational databases are designed to be accurate and amenable to support complex analytic queries. As a result, a majority of modern management reporting systems—that critically require accuracy in information—use relational database management systems. Business intelligence systems used by sales, finance, and marketing teams to analyze historical purchase patterns, and make forecasts for the next quarter or year, also rely on relational database management systems. The fact that these relational systems use structured data means what there is a lot of discipline that is attached to how and what data is stored. As a result, accuracy in reporting benefits.

But using all this structure to support accuracy comes with a price. And the price is that as the data volume increases, the performance of the queries that are used to retrieve the data slows down. Further, because the structured data uses pre-defined formats for what types of data can be stored, it becomes an issue when you do not know beforehand the type of data you will be getting.

21. Piper, Ben; Clinton, David. *AWS Certified Cloud Practitioner Study Guide*. Wiley. Kindle Edition, p. 159-160.

Let us take an example.

If you are capturing customer and order information, it is important to specify what comprises a customer and an order. For instance, you can specify that all customers must have a first name, last name, a shipping address, and a billing address. Similarly, for the order data entity, the data needs to specify product numbers, customer identifier for who made the purchase, the quantity purchased, item level, and the total price. Accurately capturing this information is critical so that, for instance, the company can provide accurate sales revenue views to its shareholders, and the tax authorities, as a public or a private company. Such information that feeds financial reporting for public (or private purposes) cannot do without using structured data and a relational database management system (RDBMS).

Now contrast this with the data that comes from IoT devices or social media feeds. This data may contain video, audio, textual or other types of format. The point is that such incoming data comes in large volumes and is of a format whose data type is not necessarily known beforehand. Such data is called unstructured data, and it is usually stored in a different type of database that prioritizes speed and volume over accuracy and querying ability. Oracle RDBMS or Microsoft MySQL are two popular examples of structured databases. Amazon Dynamo DB is an example of a NoSQL or an unstructured database.

Why should there be a trade-off between speed and accuracy, one might ask.

The answer comes from a theorem from theoretical computer science, known as the CAP theorem. If consistency means that every data read receives the most recent update (or an error), and availability means that every data request receives a response even though it may not be the most recent, then the CAP theorem states that in the event of a network partition failure, one will have to choose[22]. Structured databases such as the RDBMS choose consistency over availability[23].

22. "CAP Theorem." Wikipedia, en.wikipedia.org/wiki/CAP_theorem. Accessed 9 Mar. 2021.

23. Ibid.

Whereas the unstructured ones, such as the NoSQL, choose availability over consistency[24].

Because they prioritize data availability, NoSQL databases can support applications that perform tens of thousands of reads or writes per second[25]. In addition, since the requirements for the incoming data to align to a pre-defined format are relatively relaxed, it means that the applications using the NoSQL or unstructured data show a lot of flexibility on what types of data they can consume. Of course, such flexibility, speed of access, and ability to support large scale comes with a price: the queries against such unstructured data, especially the ones that are required for a deeper analysis, take longer[26].

Regardless, the NoSQL databases are the innovation in technology that has allowed the digital revolution to take place. Without them, all the data from the IoT devices, the social media channels, and so on will be completely useless because it could not be stored, processed, and tapped for insights and business value.

1.8 Information Security

Technology that has made the security of data possible as it travels to and stays in places other than a company's own data centers and servers is known as encryption.

In encryption, data in plain text is encoded into an alternate form known as ciphertext that is unintelligible to humans[27]. The encoded data is accompanied by an algorithm-generated key, known as the encryption key. Only authorized systems can use the key to decode the data for processing, a step known as decryption. The authorized decoding system processes the provided information, generates a response that the system encodes again for shipping to the original system. What this entire chain of encoding information, decoding information, encoding

24. Ibid.
25. Piper, Ben; Clinton, David, ibid., p. 159-160.
26. Ibid.
27. "Encryption." Wikipedia, en.wikipedia.org/wiki/Encryption. Accessed 19 Mar. 2021.

response, decoding response—all tied to an authorizing key—does is that it prevents unauthorized access to the information is it travels over the wire and gets processed in an alien environment such as that of the cloud service provider.

Now, in principle, it is possible for an entity to decrypt the message even if that entity does not possess the authorizing key, but for that to happen, the entity would have to deploy considerable computing resources for a long time[28]. For all practical purposes, the technology of encryption and decryption has made commercial transactions take place with security on the internet and the cloud environments.

When a corporate client ships data to Amazon Cloud (AWS), the client can encrypt the data itself using what is called client-side encryption. In such a scenario, the client encrypts the data before sending it to AWS servers[29]. The client can generate its own encryption keys or do that using AWS's key provisioning services[30].

As an alternative to client-side encryption, the client can delegate encrypting responsibility to the cloud service provider. Using the example of Amazon cloud again, a client can have AWS perform encryption on the server-side (a process known as server-side encryption). For instance, the client can require AWS to encrypt the data before it is loaded to database tables so that the database tables always contain encrypted data[31]. In this scenario, AWS manages key authorized encryption and decryption of information, in an automated way, without requiring inputs or intervention from the client[32].

Cloud service providers also provide a double layer of security over access to services. Known as multi-factor authentication (MFA), a requesting user needs to provide two things to access a service: (a) a password; (b) a temporary digital token sent to a preset device such as a smartphone[33].

---

28. Ibid.
29. Piper, Ben; Clinton, David, ibid., p. 76.
30. Ibid.
31. Ibid.
32. Ibid.
33. Ibid., p. 70.

In today's hyper-connected world, a corporation's systems are exchanging data with customers; employees; B2B partners; IoT devices; Cloud service providers such as AWS, Azure; social media channels such as Facebook, Twitter; backend systems of business units within the same company and so on. Each point-to-point interaction is a case of a potential security vulnerability—encryption and MFA help with securing information over this universe.

But what encryption and MFA do not help with is encroachment from other tenants who are sharing the same Cloud provisioned infrastructure. Recall that what cloud service providers such as AWS, Azure, and Google Cloud do is procure massive amounts of servers, storage, and network capacity in their data centers deployed all over the globe. They then use the software called virtualization to let many companies—the tenants—use this infrastructure. The virtualization affords scaling. Thousands of users from hundreds of companies could be accessing the same infrastructure resources. This sharing of the cloud backend infrastructure among tenants obviously creates another point of security vulnerability.

Cloud service providers such as Amazon provide a feature called Virtual Private Cloud (VPC), which is a portion of the cloud network that is logically isolated from other networks on the Cloud. Traffic to and from the VPC is restricted based on user-defined access rules[34]. The VPC is actually further divided into subnets that allow for traffic control at a more granular level[35]. For instance, using subnets, both internet-facing web servers and intermediate application servers could be hosted on the same VPC, but only web servers are provided access over the internet[36]. This ensures that all tenants resident on the same Cloud play nicely with each other.

In a big way, interacting with Cloud and being able to use it in a secure way is a case of shared responsibility. There are some things that the client needs to ensure in terms of security, while there are some

34. Ibid., p. 175.
35. Ibid., p. 177.
36. Ibid.

items that the cloud service provider—because it is being paid for it—needs to ascertain.

Amazon has formalized this case of dual responsibility. In the most general sense, the security of the Cloud is deemed as Amazon's responsibility, while the security of what is in the Cloud is deemed as the responsibility of the customer[37]. Thus securing the Cloud's hardware, its network, physical locations are the responsibility of Amazon. While securing the application code or information shared with and by the applications may well be the responsibility of the paying customer.

The prospect of letting your company's precious data travel across your company's boundaries seems scary at first. Indeed this was one of the concerns that impacted the initial widespread adoption of the cloud paradigm. Now, though, these fears appear to have mostly subsided. And the adoption of Cloud now covers private and public companies of all sizes, even federal and state governments. An indicator of this widespread use perhaps can be the revenue growth of the cloud service providers.

AWS revenues grew from $1.5 billion in 2012, to $46 billion in 2020—a CAGR of 53.4%[38]. Microsoft's Azure revenue numbers are a little harder to glean, but by one estimate, in 2015, Microsoft's annualized commercial cloud revenues were $8 billion[39]; in 2020, Microsoft reported commercial cloud revenues of $59.5B[40]; implying a CAGR of 49.4%. Even though the analyzed time horizons are different, Microsoft cloud's growth rate seems comparable to that of Amazon cloud and very strong.

---

37. Ibid., p. 59.
38. Amazon Web Services. Wikipedia, en.wikipedia.org/wiki/Amazon_Web_Services. Accessed 19 Mar. 2021.
39. "Azure Market Share: Revenue, Growth & Competition (2021)." Kinsta.Com, kinsta.com/azure-market-share. Accessed 19 Mar. 2021.
40. Evans, Bob. "Microsoft Wallops Amazon in 2020 Cloud Revenue; Bigger than AWS and Google Cloud Combined." Cloudwars.Co, 8 Feb. 2021, cloudwars.co/microsoft/microsoft-wallops-amazon-in-2020-cloud-revenue-aws-google-cloud.

At this point, we are familiar with the core components of technology that act as the foundational building blocks of digital.

But a public company with upwards of say $25 billion in revenues has a myriad of concerns that substantially impact how—and if—it can use the power afforded by the foundational building blocks of digital.

For one, this company already has an existing system infrastructure that is supporting its existing revenue base and operations. This company has an existing workforce that is deployed across the globe and has to adhere to internal and external regulatory policies and compliance. There are shareholders who scour each quarterly filing for any risk to customer retention, regulatory risk exposure, revenue growth rates, and margin pressures. Executive officers of such companies can understandably take some time before they consider options such as Cloud, IoT, artificial intelligence, Big Data analytics, and so on.

These officers have to ensure that internally—within their companies—they are organized well to mitigate the risk that accompanies the new technology components' adoption and to fully unleash their power for both revenue growth and cost optimization.

To create this readiness, two categories of organization are key. The first organization category focuses on how a company's systems and applications are architected—designed, modeled, and delivered; how company's information platform interfaces with the external world—its B2B partners, third parties, and even the Cloud service providers; what abilities the company has to scale its systems at the application unit, and the subunit level.

The second organization category focuses on how well the company's people and processes are organized to reduce time to market; to allow the company to learn continuously from the market to develop commercial options for itself while mitigating the risk presented by the uncertainty in market demand; to minimize waste, and; to scale options once they have proven to be commercially viable.

We take a look at each of these categories, and their constituent concepts, next.

Building Block Two: Organization Models (Technology and Software Design)

To create value from their digital investments, modern organizations need to respond to business change, integrate well both internally within their organizations and with the external world such as B2B partners and third-party data sources, deploy for the internet and be able to scale up or down with user demand. Over time, the technology industry has learned valuable lessons that inform how technology should be best organized to address these requirements. We look at those learnings in this section.

2.1 Technology Organization for Business Change—Service-Oriented Architecture

Investments in all the building blocks of digital such as Cloud, IoT, Big Data, Artificial Intelligence amount to nothing unless the investing business can gather some returns from them. These returns are realized through the portions of business functionality that are enabled by the underlying technology. Known as IT applications or digital applications (or digital products), these atomic units combine to create business value from investments in all things technology. The success of these applications, measured by their technical and functional accuracy, and the scope of the underlying business context that they enrich, directly contributes to the ROI from technology investments.

But managing these applications is no trivial task. For one thing, their number can be huge. A consumer company with ten billion dollars in annual revenue that this author served had upwards of 8,000 such applications. Lack of discipline in management had led to the generation of waste in the application portfolio. A substantial number of these applications imitated each other in functionality, leading to an inefficient allocation of hardware, software, and human resources. It did not help that this company had grown through acquisitions: the new application portfolios that were acquired were simply added on without going through a rationalization exercise that would have weeded the duplicate ones out.

Accumulation of waste in application portfolios is a very common problem in large enterprises. These enterprises, when they embarked on their journey of information technology, put their core operations and processes on mainframe computers. These legacy systems, even though they worked well, had one problem: they were difficult to work with. When the business needs changed—as they always do—it was very difficult to change these systems because, like the monolith in the movie Space Odyssey, they came as a giant block of functionality that was opaque and that carried the risk of breaking multiple elements of functionality if you just changed one part.

Due to these integration challenges, these legacy systems also presented resistance to the new technology. Businesses were finding it hard to take advantage of the newer, lighter weight, more scalable digital technologies because the businesses' operations and funding were trapped in these inextensible legacy systems.

Over time, new organization models for organizing technology investments emerged that allowed for accommodating newer technology that can integrate with the old legacy systems. These models made it easier to model business processes while facilitating integration with the processes' underlying (and existing) technology applications.

One such popular model is that of Service Oriented Architecture, SOA for short.

What this model does is introduce a layer of abstraction over an underlying system functionality. This layer of abstraction—known as a service—provides a clean interface for how the "hidden" system functionality can be invoked. The internal operations of the service—the underlying system functionality; the internal details of how the service carries out a request—are hidden or abstracted from the outside world. This way, how the service is consumed is cleanly separated from its internal workings. A natural advantage of this approach is that the inner workings of the service can be changed—something that happens all the time in the world of IT—without requiring the service consumers to change.

The SOA model encourages distinct components of business functionality to be their own individual services. Cohesive elements of business functionality are grouped together into a service, and the functionality offered by the service is exposed for consumption through a well-defined interface known as the Application Programming Interface or API. Any consumer interested in using the service can invoke the service's API to consume its functionality. The service's guts or its core operations can be overhauled without requiring the service consumers to change since the service API is kept fixed. This can be a tremendous advantage if the service is used directly in business-to-business interactions—by keeping the service API used by the B2B consumers fixed, the downstream ripple effects for the B2B consumers can be avoided, if the service's internal design undergoes change.

The new thinking that SOA additionally promotes is that the design of a service veers away from fat to lean.

To see this, consider the following scenario.

Suppose you have a single application that offers three sets of business functionality: A, B, and C. These three sets come in a package. You get all the three, or none—not a very flexible scenario for business consumption.

Now, suppose that instead of offering this single application, you offered three smaller services, each with individual functionality for A, B, and C, respectively. In this new approach, you have given the business more choice. Instead of having to go for all the three functionality sets bundled together, the business can now choose to go only with A+B, or A+C, or B+C, in addition to going for A+B+C. You have also given business the option of going with only A, B, or C. That is, you have expanded the choice for business sevenfold: from just a single choice of only A+B+C to the just listed seven new options.

This menu of choices in terms of services for business to choose from, it turns out, can offer the business a lot of flexibility, and can help with the speed with which a given portfolio of technology applications and investments can respond to business change. Since business needs are always changing, this agility introduced in the technology and application portfolio can prove invaluable to sustain existing—and create new—sources of competitive advantage.

It might be useful to draw a distinction between an application and a service. In principle, an application is any deployment of technology to achieve a purpose. Thus, an Excel file is an application. So is an email application, and a fitness application that one uses on her mobile to log fitness activity data.

A service is a bit different because it almost always has a fulfillment aspect in response to a request. This request fulfillment often requires the service to access some backend infrastructure or data resources. This backend access is optimized through a judicious design of the service. Indeed, service design—as manifest through the SOA—reflects a maturity in thinking in the world of information technology. It leverages the painful lessons learned over the years around how to minimize waste in technology and software design and how best to provide an ability in design to accommodate business change.

As broad and diverse as the world of information technology is (recent public estimates tag a value of $3.8 trillion to worldwide IT spending in 2021[41]), all business transactions supported by systems boil down to two things: Data; and Operations on that data. Further, these operations are one of: (a) get data; (b) update data; (c) insert new data; (d) delete existing data. The concept of service design recognizes this simplicity.

The service design also recognizes that a service can be business context-specific or business context-neutral[42]. Every enterprise-scale architecture has to provide robustness, resilience to failure, the ability to track technical performance, and response time. As a result, every

---

41. "Gartner Says Worldwide IT Spending to Grow 4% in 2021." Gartner, 20 Oct. 2020, www.gartner.com/en/newsroom/press-releases/2020-10-20-gartner-says-worldwide-it-spending-to-grow-4-percent-in-2021.
42. Erl, Thomas. *Service-Oriented Architecture* (The Pearson Service Technology Series from Thomas Erl). Pearson Education. Kindle Edition, p. 23.

service developed in an enterprise has to support features for logging, tracking, error handling, and so on. These services clearly are business-context neutral—they are required for all business process contexts—and can thus be clubbed as utilitarian services[43].

Other services represent data entities such as a customer, an invoice, an account, a purchase order, an employee, etc., and thus even though relevant for business processes, they are not specific to any one business process but can serve multiple business processes. This category of services can be clubbed as entity services[44].

The last two categories are business context-specific, and they map directly to a specific business process. One of these—Erl Thomas, in his book on Service Oriented Architecture, calls them Task services—maps to the business process at the macro, parent process level. These Task services can model the workflow orchestration required for a business process that traverses multiple individual process steps. The second of these business context-specific services maps to the specific components of an overall business process. Called Microservices, these services, can be fine-tuned for performance and scale at the individual level and require special design and deployment considerations[45].

Between the cross-business process set of services—the utilitarian, and the entity services—and the business process specific set—the Task and the Microservices—what Service Oriented software architecture has provided is a la carte option to business. At the start of a business process modeling exercise, and before doing any development, business process modelers can take a look at the existing inventory of services in an enterprise to see which of the available services they can reuse and build upon.

As an example, a service to calculate total revenues (operation) for an annual report (Task service) can use entity services for commissions, accounts payable, customers, and invoices, while using utilitarian services for request authentication and data translation throughout[46].

---

43. Ibid., p. 119.
44. Ibid., p. 120.
45. Ibid., p. 124.
46. Ibid., p. 81.

A powerful echo of not reinventing the wheel, the SOA can have major impacts on the reduction of waste in an enterprise while creating a strong ability to respond to change.

2.2 Technology Organization for Integration—Enterprise Service Bus and iPaaS

Services are rationalized components offering business functionality that clearly specify the operations they will support and the data those operations will use as the currency for exchange.

From this, it becomes clear that services in action imply an exchange of data. Now, the exchange of data between services comes with its own issues. For one, there is the data format. A consumer of the service has to be able to send the request to the service in a data format that the fulfilling service supports. Then, the fulfilling service's response has to be translated, if needed, to a format that the consuming service can understand. Further, there is a transport protocol that acts as the technical transmission mechanism for the carrying of data over the wire, and there are a few transport protocols available such as the HTTP, Messaging Queues, that can be used to transport the data. A given service may support only a select combination of data format and transportation protocols, and the requesting service has to align to this combination.

In other words, when a service is deployed in the real world, two challenges immediately arise: the challenge of number, and the challenge of variety.

The challenge of number comes from the simple math of the number of combinations (interactions) that have to be supported for a given volume of services. So, the number of service-to-service interactions that have to be supported for an inventory of ten services is forty-five; the number of interactions becomes 105 when this initial service volume increases only by 5. Very soon, this number becomes unwieldy.

There is another aspect to the challenge of number: the impact of change. A small change in the API of one service has the potential to break a number of other services that interact with it.

The challenge of variety comes from the data format and protocol aspects of service interactions that we just discussed. A given service supports only a specific set of data formats and protocols. Supporting multiple service-to-service interactions in an enterprise means that this intelligence of which service supports which formats and protocols has to be embedded in multiple services. This can be a maintenance nightmare.

To remedy these challenges, the modern organization of technology uses an architectural component called the Enterprise Service Bus, or ESB.

ESBs serve as a centralized component that takes responsibility for brokering service to service interactions. They are the message intermediary. In this model, a requesting service or a client does not communicate directly with the fulfilling service. It contacts the ESB instead. The requesting service forwards its request to the ESB that then contacts the backend-interacting service; collects the backend service's response; and translates the response to a data format, if needed, that the original requesting service understands.

Thus, an ESB becomes the centralized message broker in an enterprise that takes responsibility for all message routing, message translation from one format to another, and even protocol translation from one type to another.

What this does is greatly simplify how interactions among services and even other applications take place. Further, now a given service can be modified, replaced, or even removed while the impacts on the consuming clients can be controlled or minimized. To see this, consider a simple scenario. Say a client A uses Service B that uses a third party vendor. This vendor no longer supports Service B or, worse, say, has gone bankrupt. If the ESB is used, the ESB can honor the existing client interfaces and can indeed be used to provide a stub or replacement functionality, minimizing the impacts on the service's clients.

Thus, while the Service Oriented Architecture (SOA) through a rationalized inventory of services provides an enterprise the ability to respond to business change with speed, the SOA needs an integration backbone as offered by the ESB to connect to the other services and applications in the enterprise, and with even the external Cloud, or the B2B partner's backend systems.

Now, integration can also be done on the Cloud. An emerging platform for doing this called Integration Platform as a Service, or the iPaaS, these days can provide the infrastructure required for creating system-to-system connections and for deploying services or applications on the Cloud. This can be convenient for: integrating a corporate's on-premise systems with its cloud footprint; bi-directional communications with IoT devices, mobile applications; connecting with B2B partners at scale, and so on. The benefits of iPaaS are similar to those offered by any cloud consideration: outsourced management of non-differentiating activities such as management of infrastructure, message security, adherence to compliance, and automatic sizing for scale.

Does it mean that iPaaS is a replacement for ESBs in the world of integration architecture? Probably not. ESBs can still be the preferred choice for on-premise deployment that focuses on integration with internal legacy systems and services, with their heightened need for security. iPaaS can be a useful choice for external facing integrations, for example, with Cloud, IoT, B2B partners, and mobile devices.

2.3 Technology Organization for Web—SOAP and RESTful Web Services

Both Service Oriented Architecture and Enterprise Service Buses are design ideas for organizing software well. In one case, the focus is on better business scalability (SOA); in another, the focus is on integration scalability (ESB). Together, they provide a good design framework for an enterprise to deploy its technology investments for scale and agility required to respond to business change.

However, they do not provide a clear direction on how the services should be implemented.

There are multiple choices that are available to implement these services. And a couple of these choices have emerged as the dominant ones that are best suited to address the challenges and opportunities presented by digital.

It is helpful to define two dimensions to assess these choices: Ability for distribution, and Interoperability.

Any implementation of a modern service has to be able to communicate, at the plumbing level, with the outside world, such as the systems of B2B partners, IoT environments, other cloud environments, and so on. As recent as a few years ago, this ability to operate in a distributed environment was addressed by using the concept of distributed objects. In this model, an Object represented a set of Data, and Operations that could be performed on that data. These objects could then be distributed over different locations. A client in one location, interested in using the services of an object in a remote location, could leverage remote communication technologies such as CORBA, DCOM, Java Remote Method Invocation (RMI), and .NET Remote technology to create a binary library called the Remote Proxy, and use this proxy to communicate to the remote object[47]. This approach worked well if both the client and the distributed object used the same technology (the problem of interoperability)[48]. This distributed object model also presented scalability issues: the client stuck around till the remote object responded; the server hosting the remote object retained its session memory till it had the assurance that the client session had ended, which in some cases never arrived[49].

Over time, the web arrived. And with it, a communication protocol became prevalent that was easy to use across a large distributed network. Known as the now famous, Hypertext Transfer Protocol, or HTTP, this transfer mechanism, in addition to allowing systems to talk to each other over a vast network spread out globally, also afforded the ability for communications between applications deployed on a mobile, a desktop PC, or over the internet[50]. Further, if the system

47. Daigneau Robert. *Service Design Patterns* (Addison-Wesley Signature Series (Fowler)). Pearson Education. Kindle Edition. Kindle Location 386-390 of 6739.
48. Ibid., Kindle Location 390-411 of 6739.
49. Ibid.
50. Ibid., Kindle Location 413 of 6739.

requesting a service, and the system providing the service could be designed to an agreement on how the messages will be exchanged, interpreted, and how the backend fulfilling services could be invoked, then we—through the HTTP protocol, and this new agreement—have a framework that offers the ability for distribution, and interoperability (because all the consuming and providing systems follow the same agreement). It so happens that implementing this framework also is relatively simple, contributing to its popularity.

This framework is called the Web Service framework, and it is one way of implementing a service. Other ways of implementing a service are CORBA and DCOM, but for the reasons noted above, they are not preferred for modern digital deployments[51].

The Web Service framework comes in two popular flavors; each of these flavors uses the HTPP protocol, which is the signature protocol for communication over the web. But each flavor uses the protocol a little differently.

To understand the difference between the two flavors, consider the key ingredients required for an effective communication between a requester of a service (the client), and the fulfilling service (the web service located on a remote webserver). First, there has to be a request message that specifies what is requested. Then, there has to be a clear indication for the fulfilling service for what type of backend procedure to execute to fulfill the request, and generate a response. Finally, there has to be a response message that the requesting client can understand.

In the first flavor of a service implementation as a web service, the message format is standardized to a structured message format called SOAP (Simple Object Access Protocol). The clients structure their message to align to a SOAP format and send it to a location over the web that is the address of the resource that will do the fulfillment (the resource location is identified by an address called the Uniform Resource Identifier or URI). Once the message arrives at the resource location on the web server, a web service ("the fulfilling backend web service") inspects the message, identifies the right backend procedure

51. Ibid., Kindle Location 347 of 6739.

to execute, and executes it[52]. Notice how the web service provides a consistent (service) interface to the outside world: The protocol over which the data arrives is HTTP in all cases, and; the message format is structured to a common design (SOAP). Thus, it does not matter what specific platform the backend fulfillment engine sits on. Because the exposition of this backend is made consistent through the use of the web service, it promotes interoperability across different platforms in a big way. The SOAP model implementation of a web service is similar to how the traditional postal mail services operate.

The second flavor of implementation of a service as a web service manipulates some inherent characteristics of HTTP to reduce the amount of data that has to be shipped over the wire through request or response messages. HTTP as a standard protocol for message communication supports standard operations on data. It has methods to retrieve information (GET method of HTTP); to update or create information (PUT and POST methods of HTTP); to delete information (DELETE method). The second flavor, called a RESTful web service (REST for Representational State Transfer), exploits these inherent methods of HTTP to identify and invoke the appropriate backend procedures and return the correct response to the client. The resources on the backend web server—which can be a media file, for instance, an image, or a video or an audio file; or a text file; or sets of data; or a downloadable program—are identified by a URI, and the HTTP method is used to retrieve or modify the backend stored information[53]. RESTful interactions over the web, relative to SOAP interactions, are lighter because it is not necessary to carry and ship the entire data structure graph each time. However, SOAP-based implementations of a web service could be the preferred choice where the focus on message format structure, and some development discipline, is important.

SOAP-based and especially the RESTful web services (that are relatively easier to develop and put a lighter strain on the bandwidth) have mitigated the concerns around the ability to distribute services over a

52. Ibid., Kindle Location 786 of 6739.
53. Ibid., Kindle Location 986 of 6739.

wide location, and to build them such that they can work with platforms of all types, in a big way. There is perhaps a reason that Amazon has named its entire multi-billion dollar cloud business as AWS, which stands for Amazon Web Services. A majority, if not all, of Amazon's cloud-based services, are exposed as web services.

2.4 Technology Organization for Deployment at Scale—Microservices and Containers

We learned earlier that monolithic applications such as on mainframes limit extension and the ability to change systems to accommodate new or evolved business needs—the concept of services mitigated this concern to a large extent. One can define services at various levels of granularity and then compose them to model complex business processes.

The design flexibility offered by the Service Oriented Architecture allows one to design software that is positioned to absorb business change. But this still presents a gap when it comes to actually reducing the time to market for a software release. This gap exists because while SOA helps provide the design flexibility in software, we are still missing the deployment flexibility.

All software once designed, and tested, needs to be deployed to a production environment where the software can perform against market and customer reality. Thus, in order for a customer retention service to actually reduce customer attrition for a business, the underlying software code has to be deployed to a production environment that has the hardware, software, network, storage capacity, and other runtime support to run the software.

It used to be that the software for all the services and their related components was deployed to a production environment as a single deployment bundle[54]. This presents a dilemma: All the care that was exercised to *design* the software in a non-monolithic way gets compro-

---

54. Erl, Thomas. *Service-Oriented Architecture* (The Pearson Service Technology Series from Thomas Erl). Pearson Education. Kindle Edition, p. 349.

mised because all the created software is *deployed* in a monolithic way (typical software development cycle follows the phases of planning, design, build, testing, and deployment; post-deployment software is live on production servers). Kicking the monolithic can from design to deployment obviously limits value creation. Now, because all the services are deployed together, making a change to only one service component necessitates the redeployment of the entire package, creating a wastage of effort and resources. Further, this all-or-nothing approach to deployment means that if a given service component in the deployed bundle has peculiar needs—for instance, for scaling to improve performance—the ability to do so is limited since the tuning of the environment is done at the package level and not at the individual component level. This limits the ability to scale.

Two advances have mitigated these challenges. And they are the concepts of Microservices and Container-based deployment.

In a nutshell, Microservices are micro services—they are services that perform a small or a single (micro) function such as payment, accounting, shipping, etc., and that carry all their systemic dependencies with them so that the entire package (service and its dependencies) can be deployed and managed as a single software unit[55]. According to one account, in the early 2000s, Amazon's retail website behaved like a single monolithic application, presenting scaling issues[56]. Amazon then embarked on a service-oriented architecture journey and factored out components from the monolithic code case[57]. These factored out components represented granular functionality—subsequently called Microservices—that on deployment could operate independently of other components[58]. Transitioning to Microservices has helped Amazon support the scale of its operations on its website[59].

55. H, Jeremy. "4 Microservices Examples: Amazon, Netflix, Uber, and Etsy." *DreamFactory*, 9 Mar. 2020, blog.dreamfactory.com/microservices-examples.
56. Ibid.
57. Ibid.
58. Ibid.
59. Ibid.

Netflix and Uber are other famous examples of companies that used Microservices architecture to support the scale of their operations on Cloud[60].

(Microservices development—to decrease microservices' operational footprint—prefers HTTP/REST-based design (versus HTTP/SOAP) that relatively carries lighter data payloads and asynchronous inter-service communications (to, for instance, avoid the scenario where the calling service has to wait for the called service to respond)[61]. All this promotes an independent and scale-enabling working of a deployed Microservice.)

These days packaging of Microservices and their deployment can be highly automated[62]. And this is where the second technology advance: Container-based deployment kicks in.

Containers are similar to Virtual Machines. Recall that Virtual Machines virtualize the underlying hardware—compute, memory, storage, networking—resources. This virtualization allows the same set of physical hardware to be used for multiple software tenants. This is what makes the Cloud provide value since thousands of software tenants could be running on the same hardware, reducing costs of operations. Both containers and virtual machines allow one to package a service along with all the used libraries and dependencies and provide an isolated environment to run this package[63]. What Containers do is take the idea of virtualization a step further: Instead of virtualizing the hardware stack as Virtual Machines do, Containers virtualize at the operating system level, with multiple containers that can run on top of the operating system shell[64]. What this leads to is a much lighter environment footprint for Containers relative to Virtual Machines[65].

---

60. Ibid.
61. Erl, Thomas, ibid., p. 349.
62. Ibid.
63. "What Are Containers and Their Benefits." Google Cloud, cloud.google.com/containers. Accessed 25 Mar. 2021.
64. Ibid.
65. Ibid.

Further, there are commercial grade Container management systems such as Dockers, Kubernetes from Google, that in addition to providing a sandboxed run time environment to the deployed Microservice, provide features for automated service rollout, health monitoring, and perhaps most importantly, automated scaling[66].

With the combination of Microservices and Container-based deployments, the separation from a monolithic type of design and deployment has increased. Together, these two technologies allow for granular services to be designed and then deployed as individual independent units that can scale with demand, managed through automation, and run on all types of different platforms, affording true independence from a platform type. A container-based Microservice developed once can be run on a developer's personal laptop, her company's on-premise servers, on a cloud, or on a third-party system without having to make any code changes[67]. A given Microservice can be individually scaled up or down for performance without impacting the other services in its deployed business ecosystem. Such affordability of scale is, as we have learned, what makes modern digital businesses such as Amazon, Netflix, and Uber possible.

---

66. Ibid.
67. Ibid.

Building Block Three: Organization Models (People, Process and Management)

The first two building blocks of digital that we have considered so far allow companies to tap into the power of Cloud, Big Data analytics, artificial intelligence, and so on and then create technology organization through the use of service-oriented architectures, ESBs, Microservices, container-based deployments, and web services to create a solid foundation for the use of digital.

But this foundation—which is really centered around technology thus far—needs to be augmented with processes, people organization, and management models to develop capabilities for business use. Obviously, without the use of the right people organization and management models, all the technology building blocks cannot be brought to use to create value. These people and management-centric models are what we consider now in Building Block Three.

As with the earlier two building blocks, there are challenges specific to how people and teams must be organized and managed that digital needs to address. To see these specific challenges, we start with the markets.

When we considered the demand patterns for digital, we saw how customer experience excellence is a major requirement for companies to be competitive, or even viable, today. The barriers to entry to a vertical are not that strong in modern times, and increasingly, more and more of corporate growth is coming from a wise and strategic use of technology. There is a strong emerging consensus around today's business environment that is perhaps best outlined by a quote from Christopher Little, a software executive and one of the earliest chroniclers of the DevOps movement: "Every company is a technology company [these days], regardless of what business they think they're in. A bank is just an IT company with a banking license"[68]. Compounding the issue is an emergence of a strong consumer demographic—the Gen Y and the Gen Z—that, as savvy users of all things digital, have set a high bar of expectations from the modern brand. These powerful and vocal users not only provide a goal post of target customer experiences that the brand needs to continuously provide, but through the use of social media, these user segments also provide product design signals to the brand. The market dynamics carry a high degree of uncertainty, for it is not clear if the early signals of demand will sustain. Since the consumers' mindsets are fickle and their attention spans small, this likely contributes to the pressure of speed with which the brand needs to respond to the early signals of market demand. Specifically, the businesses no longer have the luxury to conduct strategy assessment exercises lasting months to qualify potential market demand. Businesses—to be viable today—have to: continuously explore new opportunities in the market and customer experience arenas; quickly test these opportunities for commercial viability; quickly scale these market-validated opportunities for widespread commercial use, if the tests prove successful; learn from the failures and house the learnings to make future tests successful, if the tests fail. This entire cycle which is based on scientific thinking (hypotheses are developed that are then tested in the market before acceptance and scale-up), needs to be run on an industrial scale—with speed.

Developing software that addresses these powerful market trends requires the use of four core capabilities. Building software using these capabilities is usually termed Product Engineering, or Digital Product Engineering, or just Digital Engineering (the term Engineering appears because perhaps, as we will soon see, the used capabilities leverage insights from the manufacturing industry, and from companies such as Toyota).

Use of these capabilities of Product Engineering, a company increases the chance that the software—or equivalently the digital products—it

68. Kim, Gene; Humble, Jez; Debois, Patrick; Willis, John (2016-02-07T01:28:15). *The DevOps Handbook: How to Create World-Class Agility, Reliability, and Security in Technology Organizations* (Kindle Locations 661-663). IT Revolution Press. Kindle Edition.

creates to address the market demand are accurate (the products effectively serve customer needs), their development is fast (reduced time to market), and once the product's commercial viability is established through experimentation success, the product scale-up can address demand volume and elasticity.

These four capabilities of Product Engineering are:

1. **Cloud-native development:** A disciplined approach to software application or service (or equivalently, software product) design and development that increases the factoring out of undifferentiated and non-value-adding tasks from the development of applications. Increase of such factoring further adds to the time that application developers have to focus on creative work. As we shall soon see, cloud-native development is more than just putting applications on the Cloud.
2. **Design thinking:** A structured and systemic approach to involving customers—early—in the design of software products. Done right, this approach can uncover customer needs that even the customers themselves are not aware of. Continuously uncovering and addressing unmet customer needs can lead to the design of appealing products and services, and drive exceptional customer experience, creating breakthrough performance.
3. **Agile-based software development:** An approach to building software that leverages lessons learned from the Manufacturing industry, Lean thinking, and other recent work management philosophies to increase well, the agility, with which a business can respond to change in external and internal conditions. By developing software in small increments, limiting work-in-progress, providing visibility to work, and continuous integration with the baseline, the speed with which accurate software gets built can increase significantly.
4. **DevOps-based culture and team formations:** A boundary-bashing mindset to creating software development and maintenance teams. By putting various skillsets such as for product ownership, development, IT operations, Quality Assurance, and Information

Security in one team, organizations can create powerful team units that can sustain software development from conception to delivery. DevOps maximizes the use of automation to support the speed of software deployment to production servers while ensuring that such deployments have quality built into them.

All these four capabilities are required to create velocity with which the development of software products takes place to meet the market or customer demand. This is especially true of the Agile and DevOps capabilities that typically needed to be welded together. To see this, it must be called out that while Agile drives software product development from ideation to code, DevOps takes this built code and supports its subsequent journey to a production environment. Without DevOps, the developed code would be like an inventory sitting in a warehouse that never makes it to a retail outlet and thus leads to no commercial returns for the business. Similarly, without cloud-native development, use of Agile and DevOps—and for that matter, even Design Thinking—will be of limited use because the application that has been built to meet the lucrative customer need, breaks easily, takes very long to fix, and cannot be scaled up to meet customer demand without paying exorbitant cloud-service provider fees.

We take a deeper look at each of these capabilities next.

3.1 Cloud-Native Development

To create an infrastructure for technology, a company needs to purchase capacity for physical servers that provide it with key requirements for processing all information: storage, processing power, and memory. Cloud technology, through the use of virtualization, allows companies to do more with the acquired physical capacity. Specifically, the Cloud allows companies to process a higher number of user transactions at a lower price, and as an operational—versus capital—expense. The acquired cloud capacity automatically scales up or down with user demand, and the company pays only for what it uses.

We learned all these advantages of Cloud in the Building Block One of Digital.

Now consider the following scenario.

A company acquires cloud capacity from a cloud service provider. However, on it, it deploys applications (we will use the terms applications, services, digital products, or products interchangeably) that are fat in functionality; the deployed applications show a high degree of coupling with other applications (that is, the number of interactions they have with other applications to do their job is high), and worse, use code specific to the cloud platform. The costs of maintaining this application—for instance, costs to resolve bugs in this application—are likely going to be high. Also, this application will be difficult to scale when the user demand goes up (a common manifestation of this issue is that end-user requests time out or have very slow response times). Finally, if down the line, the company wants to move to a different cloud platform, this application will need to be modified because it was written with code that was specific to one cloud platform.

In other words, even though the company has taken a step forward in acquiring elastically scalable capabilities through the provisioning of its infrastructure on the Cloud, the company does not really see end results because its applications are poorly written.

This is where cloud-native development comes in.

What cloud-native development does is provide a protocol with which applications must be written so as to maximize the benefits of the application's deployment on the Cloud. Further, cloud-native development formalizes all the automation and general cloud management services that need to be provided by the cloud service provider.

So, for instance, management of virtual servers, storage, databases, directory services must be done by the cloud vendor in an automated way and such that an application developer can use these services easily. The good news is that most of the major cloud vendors provide these services and much more—and provide them well.

The onus for doing cloud-native development then falls mostly on how applications are designed and developed. Here, a few factors are at play.

First, an application must be designed and developed using a set of twelve principles. An application developed this way is called a twelve-factor application. When designed and developed in this manner, the application increases its ability to interoperate with other applications in a cloud environment, can be deployed on different cloud vendor platforms with ease, scales easily, and minimizes the cost of adding new developers[69]. There are other rules augmenting these twelve. For instance, if the application is also developed as a Microservice, then it can be bundled as a cohesive package that can be independently deployed—and scaled—across different platforms. That is, such an application supports the "build once, run anywhere" goal. There are specific deployment technologies these days, called Containers, that give an individual secure home to these Microservice applications on cloud platforms and allow these applications to scale up and down with tremendous ease. One such popular container technology is called Docker, which allows one to bundle the Microservice application with all its dependencies and libraries into a clean, scale-friendly package[70]. When you have hundreds or thousands of such Microservice-application supporting containers running on a cloud platform, you need another service to manage the lifecycle and operations of all these containers. Google designed a service called Kubernetes to do exactly that: Kubernetes can manage the orchestration of Docker containers[71].

These disciplines of designing and developing an application—Twelve-factor application, Microservices, Container-based deployments—and then the use of technologies such as Dockers and Kubernetes to control and support the deployment of such applications, mean that most of the run time and deployment headache of operating an application on a cloud platform has been taken away from application developers. This allows application developers to focus their energies and talent on the creative aspects of designing an application

---

69. Laszewski, Tom; Arora, Kamal; Farr, Erik; Zonooz, Piyum. *Cloud Native Architectures: Design high-availability and cost-effective applications for the cloud.* Packt Publishing. Kindle Edition, p. 16.
70. Ibid., p. 137.
71. Ibid.

and work more closely with business as it tries to shape value from its digital investments.

All this positions the company to maximize the returns from its investments in the Cloud.

3.2 Design Thinking

One of the popular and powerful areas of value from digital is in shaping customer experiences with the brand. Experience of customers with the brand's channels—physical as well as digital such as mobile, and the web—the brand's products, and the brand's presence on the social media, define how much of brand equity builds up over time, how product differentiation takes place in the market, and how customer loyalty is sustained.

A critical factor in achieving this value is how well the brand systematically identifies and taps the unmet needs of the customers.

A recent approach has emerged as a powerful way to do this well. Called Design Thinking—the term "design thinking" was coined by David Kelley and Tim Brown of the design firm IDEO in the 1990s—the approach aspires to develop a deep understanding of the customer behavior, their experience with the brand and its products, their expectations from potentialities for future products and experiences, and combine all this understanding to build products and solutions that effectively address the unmet needs of the customers[72]. In this way, the approach applies a human-centric and customer-need-hinged mindset to product and solution design. Doing this early, while the conception of a product or solution idea is taking place, avoids expensive mistakes later on. (It was the famous architect Frank Lloyd Wright who said: "You can use an eraser on the drafting table or a sledgehammer on the construction site."[73])

72. Gibbons, Sarah. "Design Thinking 101." Nielsen Norman Group, 31 July 2016, www.nngroup.com/articles/design-thinking.
73. Turnali, Kaan. "What Is Design Thinking?" *Forbes*, 25 May 2017, www.forbes.com/sites/sap/2015/05/10/what-is-design-thinking/?sh=7757f833471f.

The idea behind Design Thinking is not new. For instance, in the 1960s, the clothing designer Jean Muir put emphasis on "how her clothes felt to wear", suggesting an early awareness of the end customer's experience in her design[74].

Even though Design Thinking's center of gravity is creativity, an important first step of the approach revolves around developing an understanding of customers, using empathy[75]. Empathy with the end customer's perspectives is achieved by directly observing their behavior and actions to develop an understanding of their emotions, motivations, thoughts, and wants[76, 77].

This understanding is then used to generate ideas for innovation. It is encouraged that these generated ideas—which are really hypotheses for value—are given a shape through a working product or an offering prototype and that this prototype is then presented to real customers to solicit their feedback[78]. During the ideation phase, it is important that the assumptions behind both the customer needs as well as the prospective solutions are challenged to avoid the incidence of false positives in either the need or the solution[79].

The approach uses scientific thinking. The hypotheses for value—the ideas—are tested through working prototypes. After the value proposition and the feasibility—both technical and economic—of the working prototype have been ascertained in the market, for example, through the field tests, the prototype is then scaled up for widespread commercial use. All throughout, the process of Design Thinking—using the steps of empathy, need definition, ideation, prototyping, and testing—is very iterative[80]. The use of iterations promotes learning by

74. Gibbons, Sarah, ibid.
75. Ibid.
76. Ibid.
77. "What Is Design Thinking and Why Is It So Popular?" The Interaction Design Foundation, 12 July 2020, www.interaction-design.org/literature/article/what-is-design-thinking-and-why-is-it-so-popular.
78. Gibbons, Sarah, ibid.
79. "What Is Design Thinking and Why Is It So Popular?", ibid.
80. Ibid.

doing while allowing the human mind to gain a slow yet thorough understanding of what is being shaped up, and why.

Design thinking has found a way to use scientific and rational thinking (use of hypothesis, test, and evidence-based approach) to capture not only the human needs and wants (the tangibles, so to speak) but also the intangibles of customers' emotions and motivations. As an example, one design team that the author had exposure to deployed anthropologist PhDs to study the behavior of families inside their homes—specifically, their use of wireless bandwidth. After observing their use for a couple of weeks, the team burst a set of 17 ideas for potential prototyping. Three of these ideas were selected for prototyping, and one made it to the final product development stage.

Developing the product innovation ideas using a customer feedback-based systematic approach, and testing the ideas early using market-based evidence, before handing the ideas over to engineers for a full-scale product or service development, minimizes the need to use a sledgehammer later on.

3.3 Agile-based Software Development

At the end of the day, software development is a manufacturing process. It is no wonder, then, that the software development process of today uses insights learned over the years by the manufacturing industry.

The car company Toyota's 1930 operating model led to the development of a Lean Manufacturing system, which in turn contributed to the development of a set of broad-based production optimization principles dubbed as Lean[81]. In 1996, James Womack and Daniel Jones defined Lean to comprise of a set of principles that allowed product companies to do more and more with less and less of human effort, equipment, time, and even space, while continuously reducing the gap with customer expectations[82]. This was achieved through continuous

81. Lean Manufacturing. Wikipedia, en.wikipedia.org/wiki/Lean_manufacturing. Accessed 19 Apr. 2021.
82. Ibid.

and incremental improvements that relentlessly identified and eliminated wasteful activities—any activity that the customer was not willing to pay for and that did not contribute to value creation was a candidate for elimination and simplification[83].

In 1997, the Lean Enterprise Institute started researching the application of Lean to other industries such as services and healthcare[84].

There were two major tenets of the Lean approach that were telling: (a) the time required to manufacture finished goods from raw materials, the manufacturing lead time, was the best predictor of the customer and employee satisfaction, and quality; (b) small batch sizes were one of the best predictors of short lead times[85].

A few years later, in 2001, seventeen software developers, the industry's leading thinkers, met at a resort in Utah and developed the Manifesto for Agile Software Development[86]. This manifesto, which was really a set of principles, prioritized working software over documentation; responding to change over following a plan; individuals and interactions over processes and tools; and customer collaboration over contract negotiation[87].

This spawned the Agile movement, which is really how the software is built in a vast majority of organizations today. The Agile approach aspires to increase the speed with which a business can respond to change when it uses technology or software. To quite an extent, the Agile way was a reaction to the older way of developing software—the Waterfall approach—that used sequential phases. In the Waterfall approach, the software requirements had to be defined and baselined first before they could be designed and built into code, which had to be tested before releasing into production for live use. The problem with this old approach was that it could be months before working software

---

83. Ibid.
84. Kim, Gene; Humble, Jez; Debois, Patrick; Willis, John (2106-02-07T01:28:15). *The DevOps Handbook: How to Create World-Class Agility, Reliability, and Security in Technology Organizations*, page 4. IT Revolution Press. Kindle Edition.
85. Ibid.
86. Agile Software Development. Wikipedia, en.wikipedia.org/wiki/Agile_software_development. Accessed 19 Apr. 2021.
87. Ibid.

became available. This was a problem in cases—and there was a majority of them—where the business environment changed during the time from requirements definition to production so that the value from the now available software was either impaired or diminished altogether. Further, Waterfall promoted the use of documentation. Documents were written to define requirements, for technical design, functional design, delivery plan, test plan, test scripts, test results, production deployment plans, deployment criteria, and so on. The extensive documentation created by the Waterfall approach was perhaps convenient for the system integrators who could tie their contract payment milestones to interim documents, but clearly, it produced waste by creating activities that produced little to no customer value.

Over time, multiple methodology flavors of Agile emerged, such as Extreme Programming (XP), Scrum, Kanban, Lean Software Development, Feature Driven Development, Dynamic Systems Development Method, and the Crystal Family[88]. The popular ones are Scrum, Kanban, Lean Software development, followed by XP.

Regardless of the specific Agile methodology being deployed, some of the common themes used to develop software are:

- **Reduce the batch size:** develop software in small chunks to create frequent releases of software functionality; use short development cycles (weeks instead of months).
- **Perform test-driven development:** Write test cases prior to writing the software code. The software code will have to pass these tests before it is taken to the next level. Automate these tests so that they can be run repetitively by requiring minimal human effort—minimizing waste.
- **Compile code regularly:** Create binary versions of the software code regularly so that it becomes easier to trace errors. Integrate the code developed in small chunks regularly with the master code repository. Run automated tests on the developed code so that

88. Ashmore Ph.D., Sondra. *Introduction to Agile Methods.* Pearson Education. Kindle Edition, p. 49.

only code that has passed the tests is integrated into the master repository and so that the integrity of the master code repository is always preserved.

- **Build quality directly at the source:** Use swift mechanisms of feedback from the downstream cycles so that quality issues can be detected—and mitigated—closer to the source.
- **Develop teams that are self-regulating:** The development teams, for the most part, manage themselves and learn collectively so that over time their performance improves.
- **Honor the "principle of meditation":** Teams that are allowed to focus on software development while tuning out the noise from other factors such as organization flux, politics, and even stakeholder management, perform better. The Scrum methodology flavor of Agile formally honors this aspect by having a role of a Scrum Master whose job it is to minimize disruptions and remove obstacles for the development team.
- **Visualize workflows:** For instance, Kanban uses a board with columns and cards that tells all the team members how many items are in the backlog queue, work-in-progress, in testing, and live—having this visual understanding of the workflow helps teams optimize the workflow[89].
- **Limit work in progress:** Also a Kanban concept, limiting the work-in-progress positions the development team for success as it matches the team's workload with the team's capacity.
- **Develop team norms:** Fostering team behavior can create improvements in quality and drive continuous improvement over the long term. Examples include the Scrum's 15-minute daily stand up meeting (where speakership is restricted to doers, and executive stakeholders, if attending, are allowed to only observe and not speak), arrival on time, Retrospective meetings that take a look at work just executed to see how things can be improved further. Also falling in this category is the XP's pair programming model, where developers program code in pairs to foster quality, learning, and

89. Ibid., p. 60-61.

mentorship (in a version that the author has seen, the paired programmers were encouraged to arrive, have lunch, and leave work at the same times; at all times during the day, the two programmers sat on a single workstation, with two monitors; one developer was coding, while the other was observing and providing active feedback).

Technically, the use of Agile leads to what is known as Continuous Build, Continuous Integration—a robust combination of work management processes, automation tools, and team dynamics—where working—and tested—software is added to the sacrosanct master code repository multiple times a day[90]. All the building of software, its testing, and deployment uses tools that maximize automation of the individual steps.

How teams are composed and how work is planned using the Agile approach are critical success factors for the realization of value—with speed. For if the team composition creates friction or does not have adequate skillsets—creating dependency on outside players—it can drastically reduce the speed of production, regardless of the Agile methodology flavor used. Similarly, Agile, whose major value proposition is that it does not require or support long cycles of planning—for instance, as done in the older Waterfall approach—a lot of time spent in planning can be counterproductive. But since work is involved, there has to be at least some level of planning in Agile. We look at these two aspects—team composition, and work planning—in Agile next. To do this, we will use the most adopted methodology flavor of Agile: Scrum.

A well-assembled team in Agile has roles representing the key dimensions of customer need, business value, technology, and throughput performance.

The first-among-equals role in a Scrum team is that of the Product Owner. This individual owns the business value conception, needs prioritization, and value realization cadence from the software product

90. Twitter, Marko Anastasov. "What's the Difference Between Continuous Integration, Continuous Deployment and Continuous Delivery?" Semaphore, 21 Apr. 2021, semaphoreci.com/blog/2017/07/27/what-is-the-difference-between-continuous-integration-continuous-deployment-and-continuous-delivery.html.

being developed. Product owners define value requirements from the software product, and once the software product is deemed ready for release, they are responsible for accepting the product.

The second most important role is that of a Scrum Master. Summarily, a Scrum Master serves as a development lead, who has the responsibility to remove roadblocks for the team[91, 92]. When the development team restarts its work after facing an interruption, there is some latency before the team hits its pre-interruption productivity; this latency is waste; the Scrum Master's job is to minimize this waste by shielding the team from these interruptions[93]. For instance, if the team has questions that require follow-ups from outside the team to get answers, or if the team needs to ensure compliance, the Scrum Master can chase the resolution of these issues for the team[94].

There can be an Agile Coach on the team who, as the name suggests, can help ensure that the team is following the Agile best practices.

In addition to these roles, there will be a few developers and QA persons (the testers) on the team. Scrum teams are not designed to be big. Typical team composition restricts itself to anywhere between five and nine people[95].

Work management has two aspects: How the work output is produced from value ideas, and how the work capacity is planned and managed.

For the first aspect, the product backlog is the suite of all the value requirements that the product needs to fulfill. The product backlog is owned and prioritized by the Product Owner. Priority items are picked from the backlog for work execution (production) by the development teams. The picked up items comprise a small work cycle called a Sprint that typically lasts 2-4 weeks and whose goal is to deliver working soft-

91. Agile Alliance. "What Is a Scrum Master?" Agile Alliance |, 26 Mar. 2021, www.agilealliance.org/glossary/scrum-master.
92. Geekbot. "The Main Responsibilities of a Scrum Master." Geekbot Blog, 17 Nov. 2020, geekbot.com/blog/the-main-responsibilities-of-a-scrum-master.
93. Ashmore Ph.D., Sondra, ibid., p. 78.
94. Agile Alliance. "What Is a Scrum Master?", ibid.
95. Ashmore Ph.D., Sondra, ibid., p. 87.

ware to the end customer at the end of these 2-4 weeks; all the required technical design, development, testing, and customer validation work is contained within the sprint[96].

The customer needs—the anchors of value—are mapped to requirements of the product—functionality that the product must deliver. In the Scrum flavor, requirements are called user stories, which must be small enough to get designed, coded, and tested within the same sprint[97]. While the Product Owner informs the priority of a user story, and thus suggests when it should be sequenced for delivery, it is the development team that has the best knowledge for whether a given user story, given its technical and development footprint, can fit inside a sprint, and thus decides when the user story gets scheduled for delivery.

User stories that are notionally too big for development, are categorized as Epics. When the time is right, Epics get picked up for triage, analysis, work estimation, and pick up by sprints for release to the customer.

In Scrum, the product backlog's individual elements are estimated in T-Shirt sizes (indicators suggestive of a high, medium, or low work complexity effort), or story points that use something like a mathematical Fibonacci sequence (1, 2, 3, 5, 8, 13, 21 and so on) to give effort indicators to a user story[98]. A key point is to not get bogged down by defining detailed or exact estimates, but to show acceptance for uncertainty in estimation. Important are relative—and not exact—effort sizes. A user story whose work estimation is greater than 13 points is perhaps too large for a sprint and needs to be broken down[99]. Over time, as the same team gets a few sprints under its belt, its accuracy and speed in estimation should increase. Also, over time, the same team's work performance—both the amount and speed of its throughput—should increase. That is, over time, if the team is following Agile practices well, the team becomes self-optimizing.

96. Ibid., p. 54.
97. Ibid., p. 113.
98. Ibid., p. 156.
99. Ibid.

3.4 DevOps-based Culture, and Team formations

The use of Agile helps with creating velocity of software development from ideation to working code. This working code, however, sits in a pre-production environment waiting for operational risk mitigating testing. Once the code passes this testing, that can, for instance, validate that the new production bound code will not break existing operations, will not create security loopholes, and will not crash the existing working systems, it makes it to the production system, and the new code is said to have gone live. While Agile can be seen as driving the continuous integration goal of taking the new code and integrating it with the master code repository trunk, the step of taking the new code from the code base trunk to a production environment and doing it on a repetitive and frequent basis—a goal called continuous delivery—is critical because without it the working code will always be sitting in a pre-production environment, and never see the light of the day by going live.

To complete the upstream picture, Design thinking, can in a way, create speed and accuracy in how customer needs are captured and converted to viable ideas for software development. Agile then takes these ideas forward and brings them all the way to the point of working code sitting in a pre-production environment. What is still missing is a catalyst to expedite the movement of this working code from a pre-production to a production environment.

This last step can be revenue loss generating and highly damaging to a business if not done right.

To see this, consider the following scenario.

Say, a team of excellent designers has figured out how to systemically deploy Design Thinking to create a solid pipeline of ideas that will each address the unmet customer need. Each of these ideas is tied to a specific number for revenue growth. Collectively, conversion of these ideas to software products and features is expected to increase the company's market share by a few hundred basis points over a few quarters.

The Agile team, comprised of a talented product owner, a scrum master, and developers, has had a few sprint cycles under its belt and is working well as a group. Over time, the duration of successive sprints is decreasing, and the team has been able to increase its throughput—measured as the number of user stories that have been converted to working code.

The problem is that the design plus software development engine, which is running on steroids, is not leading to any benefits for the business because the code that this engine produces has to wait a long time to move from pre-production to a production environment. The issue is that the IT Operations team that manages the production systems, and the migration from pre-production to production, uses manual steps and is overwhelmed from keeping pace with the production issues that keep on surfacing with regular frequency and the diligence that is required to make sure that the new code meets the criteria to move to production. The last few deployments of the new code to production have been a nightmare. One of these deployments particularly presented a security vulnerability on the website that was so severe that the site had to be brought down for a few hours to reverse the deployment of the new code. On average, the website generates around 20% of overall company revenues. The website outage led to a tangible loss of revenues for the company.

On root cause analysis, it was found that the development team did not have sufficient criteria for testing security compliance of the new working code that they built. In their exciting quest to reduce product backlog, the development team had missed these testing steps. Further, the team had taken some shortcuts when it came to following best practices for software architecture. Code reviews were infrequent to non-existent. There was no outside pair of eyes that made sure that the new code bring developed will not break existing systems in production—a responsibility typically associated with IT operations.

The scenario just described is more common than one would think. Contributing to the problem is the issue of speed. If you take three tranches of talent profiles—Design Thinking Team players, "the designers"; the Agile software developers, "the developers"; and the IT Operations, the system administrators, and the security personnel who ensure that the systems in production are running well—it is easy to see that the requirements for human creativity decrease from left to right, from designers to developers to IT Operations. On the other hand, the requirements for risk-aversion increase drastically from left to right.

The designers arrived late on the IT scene; the developers and the IT Operations staff, on the other hand, have been around since the beginning. And because of the different requirements of speed and risk-aversion, there historically has been a boundary between the two talent profiles of developers and IT Operations. Internal organizational boundaries are seldom good for the bottom line of the organization. The silos between developers and IT Operations meant that the code that the developers produced took too long for the IT Operations to absorb in production. Any new deployments resulting from this siloed approach increased the number of production issues that the IT Operations now had to grapple with. This further reduced the bandwidth that the IT Operations had to support newer requests for deployment. And a vicious cycle started: The IT Operations team that was already late in supporting the new deployments now had even less availability to support the backlog of pending deployments. As a result, the backlog increased.

The result of all this is the fairly common business perception of IT: "IT is so slow to respond to change," "We do not see IT as a partner to pursue the market opportunities," and so on. The powerful business units, in some cases, take the matter into their own hands and build a shadow IT of their own, distinct from the enterprise IT. Over the long term, of course, this leads to a bourgeoning IT overhead for the company.

The solution is natural: Make the IT Operations and Security personnel part of the development teams, and make operational quality assurance and security risk mitigation an active part of the upfront development processes—that is, shift operational and security compliance from far-right to the extreme left in the software development process. This is the crux of another movement that augments Agile: DevOps. While Agile's goal is continuous integration of the new code with the master version-controlled software trunk, DevOps goal is that of continuous delivery: Through the use of automation, well-defined processes, take the code from the trunk to production, where it goes live and can finally contribute to business value.

There is a strong culture and team aspect to DevOps, whose success requires that a team comprised of a product owner, a scrum master, developers, QA, IT Operations, and Information Security personnel collaborates and works well as a unit[100].

Both Agile and DevOps movements inform from the lessons of Lean Manufacturing and other advances in management thinking. DevOps, as a term, was coined in 2009, and when Agile and DevOps are working well, a company can see multiples of software deployments to production each day[101].

DevOps maximizes the use of automation tools for testing cycles such as regression testing to make sure older working code on production doesn't break; performance testing to ensure that the response time, traffic load requirements of the production environment do not degrade; security testing to ensure that the new code does not, for instance, make the production sites vulnerable to outside attacks and so on. Automated tools are also used for migrating the code from pre-production to a production environment and for making sure that the configuration integrity of all the environments is maintained.

Agile and DevOps really are Siamese twins for business value. They both are required to create agility for the business to change and to create velocity in software development that effectively reduces the time to market.

3.5 The Product Engineering Model

In 1999, during the dot com craze, three of the greatest personalities in sports—John Elway, Michael Jordan, and Wayne Gretzky—founded

---

100. Kim, Gene; Humble, Jez; Debois, Patrick; Willis, John (2106-02-07T01:28:15). The DevOps Handbook: How to Create World-Class Agility, Reliability, and Security in Technology Organizations (Kindle Locations 420-421). IT Revolution Press. Kindle Edition.
101. Ibid., Kindle Locations 432-433.

MVP.com to sell sporting goods online[102]. There were issues. For example, popular items such as snow ski boots and bike/ski racks were missing from the site[103]. There were some concerns around whether the eCommerce business model could really apply to sell products where customers value the look and feel of the products before they make a purchasing decision[104]. The company struggled to attract a loyal customer base, and the website experience of consumers was "clunky at best, and exasperating at worst," according to one report[105]. Even though hinged around an eCommerce business model, instead of focusing on customer experience, the company made a bunch of acquisitions, diluting its brand[106]. As a result, the company folded in 2001, and the eCommerce venture of its celebrity founders came to an end[107].

In contrast, Nick Swinmurn founded Zappos, an online shoe and clothing retailer, in 1999, using the principles of the scientific process[108]. Swinmurn wanted to test the hypothesis that customers were willing and ready to purchase shoes over the web[109]. Rather than building a website, Swinmurn took pictures of shoe inventory in local shoe stores, put the pictures over the web, and after he had made a sale, purchased the shoes at full price from the local stores, and then shipped the shoes directly to the customers[110]. This helped Swinmurn ensure that the customer demand existed for the model and perhaps also allowed him to learn a thing or two about the fulfillment process[111]. In 2015, Zappos

102. "MVP.Com Is Founded." World History Project, 1 Jan. 1999, worldhistoryproject.org/1999/mvpcom-is-founded.
103. Ibid.
104. Ibid.
105. "MVP.Com Strikes out—Feb. 7, 2001." CNNMoney, 7 Feb. 2001, money.cnn.com/2001/02/07/sunsetting/biz_mvp/index.htm.
106. Ibid.
107. Ibid.
108. Wikipedia contributors. "Zappos." Wikipedia, 26 May 2021, en.wikipedia.org/wiki/Zappos.
109. Ibid.
110. Ibid.
111. Ibid.

had revenues of $2 billion, with an employee base of over 1,500—a clear success story[112].

What Swinmurn had done was used the idea of a Minimal Viable Product, MVP for short, that allows a team to learn maximally about the needs of the customers with the least effort, using a version of the product with minimal features[113]. The goal of the MVP is to allow the entrepreneurs looking to launch their startups an opportunity to test their fundamental business hypotheses or assumptions—quickly—so that they can learn about their customers' needs, and the market, at an earliest point[114]. Because the effort has been deliberately kept to the minimally needed, the cost of a potential failure is small, allowing the entrepreneur to change course, if needed, with minimal impacts. MVP based product development is not tied to software products only, and indeed is part of a best practice approach—called the Lean Startup approach, popularized by Eric Ries in his book of similar name—that aspires to position startups for success by using short product development cycles; by testing product viability using hypothesis-based experimentation and iteration-based product releases, and; through continuous and relentless learning from the market[115].

(Elway and team were perhaps thinking of the 'most valuable player' designation when they created the website MVP.com. It is perhaps ironic that in posterity, a reason for their failure was them not using the idea of a minimal viable product, another designation for MVP.)

The Lean Startup precepts are similar to those of Lean Manufacturing, and when applied to the world of software development, these are often known as the Lean Software Development[116].

Lean Software Development is also termed as Digital Product Engineering or Product Engineering. Done well, it allows companies to learn continuously about the customer need, develop hypotheses

---

112. Ibid.
113. Wikipedia contributors. "Lean Startup." Wikipedia, 23 May 2021, en.wikipedia.org/wiki/Lean_startup.
114. Ibid.
115. Ibid.
116. Ibid.

for business value, rapidly test these hypotheses in the market using MVPs, continuously experiment with ideas in the market, and learn from failures, using design thinking, and a fast and agile technology backend comprised of cloud-native development, Agile based software development processes, and DevOps based culture and team formations.

That is, in a formula, Product Engineering equals the summation of Design Thinking, Cloud-native development, Agile, and DevOps, powered by lean startup practices.

When this Product Engineering model is working well in an organization, a few positive results start to show up. First, there is the fast movement of work from left to right—from ideation to development, and from development to operations—because the production batch size is small, work is made visible throughout its flow, manufacturing iterations are small, and the development team is encouraged to constantly do better in terms of throughput, defects, and quality[117]. Second, there is a constant and fast movement of feedback from right to left (for instance, from operations and security to development), and quality adherence is made an important goal at an as upstream a point in production as possible[118]. Third, a culture and a mindset start to prevail where team performance is prized over individual performance, product decisions are based on evidence and scientific approach (hypothesis-test validation based thinking), risk-taking is not frowned upon, and organizational learning is streamlined from both successes and failures[119].

Armed with these prized capabilities, an organization starts to behave as one organism whose heart is constantly beating and responding to customer needs. This organization is relentlessly experimenting in the market for the next set of groundbreaking product or product feature ideas. The organization has enough discipline to pivot quickly

117. Kim, Gene; Jez Humble; Patrick Debois; John Willis. The DevOps Handbook: How to Create World-Class Agility, Reliability, and Security in Technology Organizations, IT Revolution Press, 2016, pp. 11-13.

118. Ibid.

119. Ibid.

from failures. Its internal departments do not work in silos but are centered around customer needs so that the organization is systematically reducing experiential gaps in its customers' journeys. Customer attrition is likely not a problem for this organization; its customers serve as advocates for its brand.

This organization is the first one to enter markets when the opportunity shows first signs of commercial viability.

And as multiple success stories of digitally native companies have shown, this organization is poised for revolutionary—not just incremental—growth.

The Digital Capability Model

As organizations of all sizes try to capture value from digital, an often-asked question is: "Where do we start?" Or, "What sequence of prioritized digital investments should we pursue?"

It helps to have a view of an inventory of digital capabilities that need to be nurtured and built over time. Such an inventory—the Digital Capability Model—is presented in Figure 1. This model is a summary depiction of building blocks one, two, and three presented so far in this book.

Figure 1

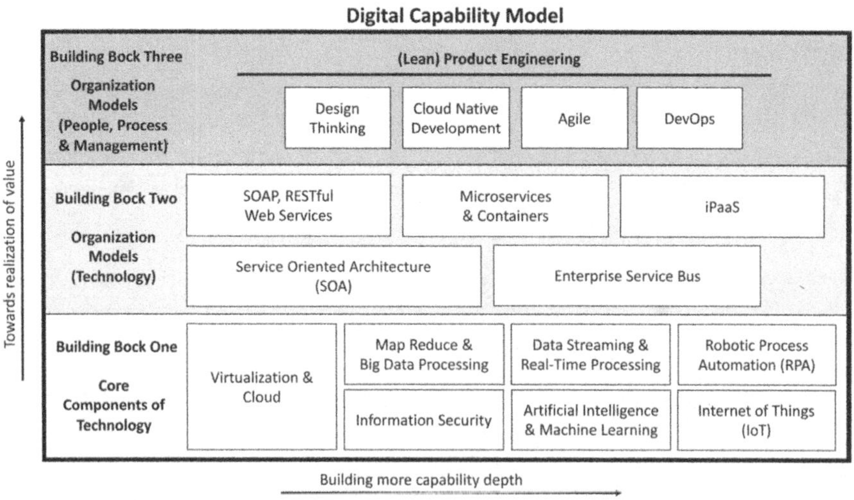

The Digital Capability Model can guide funding for the digital investment portfolio. The model captures capabilities required to build a solid digital foundation that can then be used repeatedly to produce value. When viewing the picture of the model, it might be helpful to use two dimensions to guide investment prioritization and sequencing. The horizontal dimension ("the X-axis") captures the depth and completeness of capabilities in one building block. Progress along the vertical dimension ("The Y-axis") brings an organization closer to realizing value from its digital investments. How much investment should be made to create depth and completeness of capabilities for one building block (progress on the horizontal dimension) versus investing in other building blocks as well (progress on the vertical dimension) to get closer to realizing value is a balancing act.

For instance, an organization may choose to invest in Cloud, Big Data Processing, Artificial Intelligence and Machine Learning, and Information Security from building block one; and then use its available funds to build some RESTful Web Services (building block two) through one or two pods (teams) of product engineering (building block three) so that it can start experimenting in the market. As the organization sees early success in value, and as its confidence to create value from digital increases, it can then expand its capability footprint in each building block.

Generally speaking, and as a broad guideline, it usually will be a better choice for an organization to prioritize early visibility into value. This, in turn, will require—at least initially—to build out only the most needed "horizontal" capabilities in a building block and to make sure that the organization model for technology (block two) and product engineering (block three) is activated at an early point. Horizontal capability depth can then be built in increments as realized value starts to trickle in and grow.

Sales and marketing functions are usually good candidates for building out capabilities such that value can be realized early. Both of these functions—as also the customer service function that works closely with these two—have the best proximity in an organization to the customer. These functions are responsible for shaping demand and have the best view into how a company's products are faring in the market. All this makes these functions fertile grounds of value and to build out and test digital capabilities against real-life scenarios. As an example, in these functions, Virtualization and Cloud, Big Data processing, artificial intelligence models (building block one) can be deployed with some backend services of SOA, and some front-end

RESTful web services (building block two), along with one to three pods (teams) of product engineering (building block three) that allow the company to rapidly tap the customer need and address it to create business value. In this scenario—that prioritizes rapid value realization over creating a more complete foundation—it may not be necessary to fund investments for IoT, RPA, Real-Time Processing—yet. As value starts to get realized from the digital investments in these functions, and as the entire company learns how to build digital capabilities and create value, the footprint of the foundational building blocks can be expanded.

The balancing act of how much to focus on value realization versus building out a more complete foundation differs by industry type and the company's own specific context. A company that prides itself on product differentiation and delivering exceptional customer experiences today (Retail and Consumer Packaged Goods (CPG) companies are an example) may have a stronger need to deploy a full vertical slice (that has components from *all* the three building blocks) than a manufacturing company that places far more importance on product intelligence and thus needs a robust foundation of IoT, Real-time streaming, and iPaaS before it has the confidence to go to market. The manufacturing company—relative to the retail or CPG company—likely will be building out a more complete foundation in blocks one and two before substantially scaling its investments for block three.

Similarly, heavily regulated industries such as the pharmaceutical and financial services may need to have a solid foundation as well of building blocks one and two before they can start reaping returns from their digital investments using block three. For financial services companies, especially, having a robust Security layer over Virtualization and Cloud is critical before any returns can be harvested. For the pharmaceutical companies, it helps to have a well-functioning SOA, Enterprise Service Bus architecture, and even RPA to support adherence with GxP (the industry specific good quality practice and regulatory guidelines) requirements. For pharmaceutical companies, Big Data processing and the use of artificial intelligence models have been known to reduce the R&D overhead of discovering the next promising molecule; such companies may find little use for data streaming and real-time processing for their initial funding of digital investments.

The capabilities presented in the Digital Capability Model run on data. There are some common data entities that are needed by all digital investment portfolios, regardless of the industry type. They are: Customer; Inventory; Products; Orders; Employees; and Partners.

It is critical that all these data views are holistic. That is, they need to have a 360-degree view of data covering all its possible sources.

So, for instance, a Customer 360-degree view covers the capture of customer data for an organization, across all customer touchpoints, all channels—physical and digital—and the entirety of the path to purchase journey, covering points for the brand or product awareness, interest diligence, purchase, post-purchase delivery and service, and returns.

Inventory 360 needs to cover inventory information across all locations from the supplier, through the warehouse, the distribution center, and the store. The Order 360 view needs to cover all channels—the physical store, the web, the mobile app, and the social media.

Armed with these holistic data views that these days cover not only structured but also unstructured information coming through from IoT, social media feeds, and so on, the organization can use the digital capability model to build capabilities that will generate insights for action and ultimately generate value, our focus now for the remainder of the book.

In the coming sections, we will look at how these digital capabilities can be used to create value for organizations. We will take two views: Value created by function; and value created by the industry sector.

Part Four

Value Creation from Digital

Now we take a look at how value is created from the use of digital. To do this, we will use two views. The first view takes a look at how digital can help optimize the cost of functions such as inventory or supply chain management; sales, marketing, and customer service management; finance; and human resource management. Reducing the costs of these functions contributes directly to the bottom line. In a few cases, digital, in addition to reducing the costs, can also help drive an increase in revenues. The use of digital to help drive down costs can be thought to be relatively easier than using it to drive up revenues for the simple reason of predictability: controlling internal functions to reduce costs conceivably carries a higher degree of certainty than attempting to influence or respond to external factors of customer demand, and market dynamics, which by their very nature are difficult to control, much less predict ahead of time.

The second view is about how digital contributes to value by industry type. Of course, not all industry sectors are covered. But we do use an assortment of industry types to give a fuller picture of how digital capabilities can be deployed for business use.

For companies looking to capture their first returns from their use of digital, an easy place to start is the optimization of costs. Generally, using automation to replace manual labor for mundane, transaction volume-heavy processes that have structured steps and a lower propensity for error is a good place to start.

As a close and next frontier, a company can make sure it is using structured data—capture, transformation, processing of such data, as well as generation of meaningful reports based on it—well to facilitate executive decision making. This can be closely followed by developing an ability to trap and harvest unstructured data such as that coming from social media platforms, IoT hubs, device sensors for insights. These insights continually inform ideation for improving customer experience and intelligent product design.

Value Creation from Digital, By Function

There are specific things that can be done by using digital to optimize the costs of functions, generate revenue uplift, and relieve the expense burden in an income statement. Quite a few of these have become table stakes for doing business. They are:

I. Reducing Inventory Costs
II. Optimizing Sales, Marketing and Customer Service Functions to reduce costs and generate revenues
III. Reducing costs for the Finance function
IV. Maximizing benefits from HR

We take a look at each of these individually next.

Reducing Inventory Costs

Formally, inventory as a current asset finds a spot on the balance sheet. However, inventory changes get reflected in the cost of goods sold, which is an income statement line item and directly impacts the margin of a business.

For merchandising firms, inventories are the physical goods acquired by the firm for subsequent sale. For manufacturing firms, inventories include direct and indirect raw materials, work-in-process as well as finished goods. If the inventory turnover rate is defined as the cost of goods sold over average inventory levels, then a low turnover rate may suggest overstocking, a less appealing product line, or sub-optimal marketing[1]. This can be a bleed on margin since the costs of carrying inventory—storage, service, and handling costs; taxes on inventory; capital that is stuck in the maintained levels of inventory; depreciation and insurance premium costs—can be significant[2]. Seasonal businesses, such as Retail, may face another burden from inventory obsolescence.

On the other hand, a high inventory turnover rate may indicate lower than desired inventory levels, increasing the risk of a stockout, a revenue loss[3].

Clearly, optimum levels of inventory have to be maintained across all locations—warehouse, distribution center, manufacturing floor, or retail outlets—that do not create avoidable costs for the business, and at the same time, minimize or eliminate incidences of revenue loss (through stockouts).

A desired state is just-in-time inventory where the company receives goods or materials in close proximity to when the goods or materials are actually needed.

Digital trends of modern times can help companies maintain optimum levels of inventories and mitigate lead time—the difference between when a good is ordered versus received—variability.

A powerful technological advance has been in the tracking of inventory and getting real-time visibility across the supply chain through the use of Radio-frequency identification, or RFID, tags. These tags that are attached to a specific inventory item transmit digitally encoded data to a reader over radio waves. Information collected from a sea of tags can be stored in a database for later analysis. A popular usage scenario is to use the RFID tags to track inventory or any assets as they move through the supply chain. Using RFID tags, operators can instantly track an item and use the information to proactively monitor

1. Boundless. "Reporting and Analyzing Inventories | Boundless Accounting." Lumen Learning, courses.lumenlearning.com/boundless-accounting/chapter/reporting-and-analyzing-inventories. Accessed 7 May 2021.
2. Zoho. www.zoho.com/inventory/guides/what-is-carrying-cost.html. Accessed 10 May 2021.
3. Boundless. "Reporting and Analyzing Inventories | Boundless Accounting." Ibid.

lead times with suppliers and communicate accurately with the fulfillment teams and even customers[4].

But RFID tags have been around for some time. Where the value from digital arrives is when the voluminous data collected from these tags, in addition to providing real-time visibility for tracking, is fed into artificial intelligence models that can then be trained to predict more accurate lead times and identify issues or strengths in supplier performance. Building real-time visibility into the supply chain along with the ability to predict lead times more accurately becomes a powerful combination to manage expectations with the customer, enhance customer experience with the company, improve brand perception, and of course, improve cost and revenue performance[5]. Patterns contributing to the variability in lead times can be detected to create efficiencies in ordering, receiving, and management of inventory. Past patterns can inform higher accuracy in future supply chain planning, driving better decisions for strategic sourcing and improving performance on supply chain metrics.

A supporting trend is the use of a digital platform that fosters a data-driven and transparent collaboration with suppliers to enable a just-in-time delivery model that reduces or eliminates the requirement to have safety stock, reducing the inventory carrying costs.

RFID tag induced real-time supply chain visibility, artificial intelligence model-driven better predictions in lead time and supply chain planning, and use of data and insight-driven B2B collaboration platforms with the suppliers can significantly empower a company to improve margins.

---

4. Castleford. "3 Technologies for Effective Inventory Control and Stock Management." Cohesio Group, 5 June 2019, cohesiogroup.com/news/3-technologies-for-effective-inventory-control-and-stock-management.
5. "Digital Supply Chain | TransVoyant—Part 2." Transvoyant, transvoyant.com/tag/digital-supply-chain/page/2. Accessed 7 May 2021.

Optimizing Sales, Marketing and Customer Service Functions to reduce costs and generate revenues

In most organizations, there are separate departments for the sales, marketing, and customer service functions. But it is useful to view all these three functions together when it comes to thinking about how to increase revenues and optimize costs. The reason for this is customer-centricity.

Regardless of which function is involved—sales, marketing, or customer service—the brand is dealing with the *same customer*. To drive improvement in the bottom line intra- and inter-function costs have to be reduced in such a way that the customer experience is not interrupted across touchpoints, customer intelligence is gathered from all interactions with the brand, and this intelligence is used to drive an uplift in revenues.

Typically, this involves an orchestrated management of various points in the customer journey. Early customer awareness of the brand and its products, as well as an initial generation of demand (pre-sales activities), can be marketing-led. Once a lead has been generated, its conversion to an opportunity, a quote, and eventually an order can be driven by sales. Order fulfillment and post-sale care can be managed by customer service, with coordination from sales.

Fueling the success of sales, marketing, and customer service functions is the aspect of how well the customer profile, her needs, and expectations from the brand are understood. Understanding these needs and using this understanding to develop insights that improve the performance of all the three functions are areas where digital can excel.

A good number of digital use cases that drive the performance of sales, marketing, and customer service functions can be categorized into three common threads:

A. Automation: The use of technology to automate common tasks and workflows within and across the three functions. The use of automation is mostly a cost reduction play, even though it also, as another benefit, frees up the time of human resources to focus on top-line supporting activities. An example is when sales force automation increases the amount of time a sales force personnel can spend on developing relationships with customers, driving opportunities for value-adding cross-selling and up-selling activities.
B. Analytics: The use of artificial intelligence models to predict outcomes so that proactive actions to influence favorable results can be taken; to categorize voluminous data into cohesive sets that bundle similar behavior profiles that the organization can respond to, with increasing effectiveness. Underpinning the use of analytics is a robust organizational capability—people, process, and technology—to gather data from a variety of sources such as social media feeds, sales channels, smart products, mobile devices, GPS location streams, and so on. The company's artificial intelligence engines then consume this data—that can be about customer's sentiments, their vocal perceptions about the brand, the company's products, and their ecosystem—to generate insights that are then used for—often real-time—decision making. Analytics can provide a strong contribution to top-line growth.
C. Electronic Direct to Consumer channel: Using the electronic D2C or the eCommerce or the digital commerce channel allows the brand to improve its margins by not having to share them with intermediaries such as wholesalers, resellers, or brokers because the brand sells directly to the consumer over a website. This also allows the brand to keep a higher degree of control over its marketing costs and the resulting ROI; develop a higher degree of intelligence on its customers, what they prefer, their product usage, and use this intelligence to reduce experiential gaps in customer journeys. This gathered intelligence can also be used to improve future product design. Done well, the use of eCommerce can significantly impact the generation of revenues, customer loyalty, and brand differentiation in the market.

Use of Automation in Sales, Marketing, and Customer Service functions

According to a recent McKinsey report, automation of sales-oriented tasks—covering sales strategy and planning; lead management; configuration, pricing and quoting; order management and post-sales activities—can cover a third of all sales tasks and has the potential to uplift sales by up to 10 percent, and improve efficiency by 10 to 15 percent[6].

A sales personnel's time can be freed up to spend more on managing customer relationships by automation of activities such as for reaching out to pre-identified customers, for example, through bots that can effectively drive initial engagement; generation of leads, and their assignment to the appropriate sales force personnel—for instance, by matching the lead characteristics to a sales force person's territory or area of specialty; and creation of task management workflows.

Salesforce territories, incentives, compensation, and quotas can use automated models that learn from and get better with experience, reducing the administrative overhead. Salesforce onboarding, development of job aids for the field, training, and compliance can all be automated partially or completely to reduce costs. Contract generation can be automated to the fullest extent possible, driving consistency in experience with the brand while keeping the costs of selling down.

In addition to driving down the costs for sales and marketing, automation can also reduce the costs of customer service by reducing both the volume of transactions as well as the unit cost per transaction in the customer call centers. The use of customer self-service portals, interactive voice response (IVR) can reduce the volume of inbound calls or requests to a call center. The use of intelligent automated workflows that, for instance, integrate with the back-end business process management engine and reduce the number of things that a manual agent has to do can drive down the per-unit transaction costs (falling

6. Bangia, Manu, et al. "Sales Automation: The Key to Boosting Revenue and Reducing Costs." McKinsey & Company, 26 Feb. 2021, www.mckinsey.com/business-functions/marketing-and-sales/our-insights/sales-automation-the-key-to-boosting-revenue-and-reducing-costs.

in this category are also virtual agents or chatbots that front customer requests). Automated customer service workforce management, task scheduling, tool-enabled guidance, and decision support can reduce the amount of time agents spend in servicing a transaction, driving down per unit transaction costs.

Contributing to the reduction in overall costs of customer service is also how the technician workforce that provides services in the field is managed. Similar to what helps in keeping the costs of a customer call center down, automated workforce management for the field, task scheduling, tool enabled problem triage and resolution, use of analytics to gauge technician performance and using this intelligence to make future assignments all help to minimize the overhead from field engineers. It is important that the customer call center that typically initiates field requests is integrated well with the information flows used by the field.

Use of Analytics in Sales, Marketing, and Customer Service functions

It used to be that most of a company's investments in analytics focused on creating a data warehouse that extracted streamlined data from the company's transactional systems such as for enterprise resource planning (ERP), customer relationship management (CRM), and then used this data to create reports for sales, marketing, and finance managers who used these reports to make operational decisions. A salient aspect of this approach was that the data it used involved structured data. Information could be neatly mapped into the rows and columns of a table; data entities could be represented through relational keys. Such reporting—known in the jargon as descriptive analytics—typically helped managers make operational decisions, mostly tactical even though planning and forecasting were involved. Sales forecasts, marketing budgets, finance's quarterly or yearly projections were created this way.

But of course, what all this represents is no longer sufficient to drive the performance of modern businesses. Today's companies need to augment their use of structured data with that of unstructured data: all the data that is constantly pouring in from social media platforms (about customer's sentiments, experiences with and perceptions of the brand); from smart devices and sensors (informing about product usage and ethically permissible consumer behavior with the products); customer's location as informed from the brand's app on the customer's mobile phone and so on.

Making use of this data invariably involves the use of artificial intelligence-enabled analytics, and the most powerful—and most common—of such use cases are in the realm of sales, marketing, and customer service. Broadly, these analytics use cases create value from:

1. Reducing the cost of customer acquisition
2. Reducing churn

Use of analytics to reduce the cost of customer acquisition

The customer acquisition cost is typically defined as the total marketing and sales expenditure divided by the number of acquired customers. Analytics—or Big Data Analytics to be exact—helps bring down this cost by creating superior segmentation of customers using artificial intelligence. The customer demographic and in-house operational data is enriched with external third-party provisioned data such as for customer's financial or credit profile, customer preferences, social media presence, and so on to create unsupervised learning data sets. These data sets are unsupervised because they lack data labels: there is no information for what comprises a correct result (a label). Thus, these unlabeled data sets are assessed for patterns. These patterns define the segmentation of the customer base. The machine learning models that train on these unlabeled customer data sets use clustering algorithms that find the best way to identify patterns or segment the customer base. These segments then inform specific marketing campaigns. Good and improving marketing ROI indicates the machine learning model's success.

Well-functioning deployment of such clustering models improves the quality of leads and increases the number that converts to downstream opportunities and quotes.

Use of analytics to reduce churn

Customer attrition—or churn—is a problem in every industry sector that is not governed by a monopoly. If a customer's propensity for churn—likelihood of attrition—can be predicted ahead of time, then marketing can execute outreach programs that can eliminate or reduce the probability of attrition. The performance of these outreach programs—another word is intervention campaigns—can be evaluated over time such that a most optimal intervention campaign can be launched for the next customer about to churn.

These predictive models almost invariably these days use machine learning. What such modeling does is create a data set on customer behavior with the brand, capturing customer preferences for content, channel, and time; purchasing patterns; digital commerce footprint, and so on, and then use this data set to train a machine learning model. These data sets are labeled since for each set of customer data, there is a record (result) of whether the customer churned or not. The machine learning algorithms that train on these data sets are termed classifier algorithms since they classify the assessed data into a result category: a binary answer to a question such as "will this customer churn?" And since the data is labeled, the class of learning problem that this represents is that of supervised learning.

Once a customer's propensity to churn becomes known, marketing staff can reach out to the customer with perhaps a discount to encourage retention.

Such intervention outreach from marketing can be augmented with a lifetime value calculation so that the intervention uses specificity to the customer and avoids 'a one size fits all' approach. In the lifetime value calculation also, machine learning comes into play. The customer's past purchase history and other data are used to estimate the long-term revenue potential of the customer. Past customer purchasing patterns and past scores for lifetime values can be used to create a—labeled—data set that a machine learning algorithm—such as a regression algorithm—can train on to estimate scores for future customers. Armed with the lifetime value data, marketing can go aggressive in preventing the attrition of a high lifetime value customer, while it uses a broader mass approach to prevent churn of lower lifetime value scoring customers.

Use of eCommerce to optimize sales and marketing costs, and drive customer experience

It used to be that if a company was a small brand that manufactured consumer products, and the company did not find a space on the retailers' shelves, it was hard for the company to do business; indirect channels, and intermediaries such as distributors, dictated how business was done for the company[7].

No more.

In 2013, the digital channels accounted for only 1% of all CPG (consumer packaged goods) sales; by 2018, this number had increased to 11% of the total market[8]. Millennials, as we have noted earlier, gravitate naturally towards the online channel. It turns out that with COVID-19, the attractiveness of the online channel has also been embraced by the seniors. By one account, the COVID-19 pandemic has pushed 47% of seniors—those aged 65 and above—to start shopping online[9].

Products that draw a higher use of customers' senses—taste, smell, touch, feel, sight, hearing—can be good candidates for the D2C phys-

7. Team, Oro. "Digital Transformation for CPG Industry Is Key to Success." OroCommerce, 1 Apr. 2021, oroinc.com/b2b-ecommerce/blog/digital-transformation-for-cpg-industry-key-to-success.
8. Ibid.
9. Ibid.

ical channel such as a retail store, or the D2C digital channel such as a website, or the mobile app for the simple reason that through these D2C channels, the brand can, of course, engage directly with the consumer, but in so doing can also learn extensively about what appeals to the consumer and what does not. Doing this systematically and through the use of automation can create a massive volume of intelligence about customer behavior, product usage that can then inform future undertakings for reducing experiential gaps in customer journeys and product design. Examples are WarbyParker (sells eyeglasses online), Everlane (luxe clothing), D2C in the wine industry (California's Daou Vineyards and Winery)[10].

Products that benefit from or require an intimate involvement with the consumer—one that benefits from a secluded, very personalized setting—are also good candidates for primarily, or even exclusively, D2C treatment. Examples are MeUndies (sells underwear online), AdoreMe (lingerie), Combatant Gentlemen (high quality tailored suits), Cuyana (select, high-quality closet items).

Thus, the fashion apparel and beauty CPG brands are doing well by creating either an electronic D2C only presence or by mixing it well with other D2C channels such as retail stores, social media, or even direct sales force.

Tapping into the D2C model are also other industry segments such as nutrition, home, travel, and personal care[11].

Financially, the use of the D2C channel is lucrative: It allows a business to exercise tighter control over its marketing spend, and because intermediaries are not involved, improve its profit margins.

Customer intelligence gathered through the electronic D2C channel can inform better targeting in the marketing campaigns while providing the customer a more personalized experience. Combining this channel with social media intelligence can now provide a more holis-

10. Wyrzykowski, Kajetan. "All about the D2C Business Model and How It Can Shake up Your Business." *Packhelp*, 19 May 2021, packhelp.com/d2c-direct-to-consumer.
11. "Top 133 Direct-to-Consumer (DTC) Brands & Companies." SKUBANA, www.skubana.com/blog/direct-to-consumer-brands. Accessed 12 May 2021.

tic view of the individual customer that, in addition to product usage, also has intelligence about customer's sentiments, attitudes, philosophical and political leanings, lifestyle choices, family profiles, religious beliefs—all of which can be fed into artificial intelligence models and analytics to do a better job of customer segmentation and targeting for marketing campaign effectiveness.

Direct access to the customer also enables testing of new ideas and products with the customer early in the launch cycle and at less prohibitive costs, easing the cost of failure and increasing the chance of product launch success.

But increasing the D2C in the channel mix also carries its own challenges, that if not carefully addressed, can lead to ruin.

The first and foremost of these challenges that need to be addressed is that of aligning customer experience across all channels—physical and digital.

An Accenture survey found that 78 percent of shoppers in the US had browsed online and then gone to a store to make a purchase (had "webroomed"), while 72 percent of shoppers had showroomed, that is, they had gone to a physical store first to see a product and then searched online to find a better price and had made their purchase online[12]. Aligning the customer experience across the physical store and the online channel is critical to increasing the chance that the customer will complete the purchase with the brand—on any channel. Also equally important is the requirement that the cross-channel experience is seamless—the brand makes it easier for the customer to hop across channels: physical stores as well as all the flavors of the digital channel such as the web, the mobile, and the social media.

The success of the electronic D2C channel has become table stakes for any company looking to be successful in modern times. A clear example is Amazon, predominantly an eCommerce retailer that makes

12. "More U.S. Shoppers Plan to Buy from Stores but Want the In-Store Shopping Experience to Match Convenience of Online, Accenture Study Finds." *Accenture*, 30 Jan. 2021, newsroom.accenture.com/news/more-us-shoppers-plan-to-buy-from-stores-but-want-the-in-store-shopping-experience-to-match-convenience-of-online-accenture-study-finds.htm.

billions of sales over its online and mobile sales channels. Then there are some companies that use only the electronic D2C channel; that is, they have no brick and mortar retail outlets. First Internet Bank was chartered to operate entirely online[13]. Even companies that have physical D2C channels do billions in sales over the eCommerce channel.

Optimizing the eCommerce channel has become an industry of its own. It is useful to break down the optimization of this channel into two categories:

1) Optimization of the eCommerce front end that generates traffic to the eCommerce site, analyses the traffic on the site to create customer stickiness, and then applies analytics to increase conversion rates.
2) Back-end support for eCommerce that is required to support the end-to-end customer transactions on the site.

Optimization of the eCommerce front-end

The optimization of the front end of the eCommerce channel covers both its web and its mobile manifestations. If eCommerce abilities are also offered over social media, then that exposure has to align with its web and mobile counterparts such that the customer experience across all the channels is seamless and integrated.

The front-end optimization can be further broken down into three categories:

A. **Maximizing traffic to the site:** Typically, this happens through the use of levers such as Search Engine Optimization (SEO) and the provisioning of content on the site that is relevant to the common keywords used to search the brand or its products. SEO involves developing an understanding of what users search online over search engines such as Google, and Bing, the answers the searching

13. First Internet Bank. www.firstib.com/about-us/our-people. Accessed 13 May 2021.

users are seeking, and using this understanding to influence the search results that the engine shows to the user. If a brand's eCommerce site appears prominently in the search engine results, and the user clicks on the brand's eCommerce site's link, then that is a success. Having content on the website that is relevant and credible for the user searches; is engaging and attractive to look at can help increase traffic to the site.

B. **Sustaining customer visits on the site:** Once a customer has clicked on the site's link and is on the site, the next goal is to sustain the customer at the site so that the customer advances on the path to purchase. Having an attractive product catalog that offers prospective customers on the site an ability to make product comparisons, and visualize products, helps. Understanding the intent of the prospective purchase, and then using real-time algorithms to provide product and service recommendations can motivate customers to proceed to the next step toward a purchase (these product recommendation engines in addition to ensuring that the customer has a view of product choices available to execute on her intent and complete the purchase, can also cross-sell and upsell other products to improve company revenues). Personalization of content on the site to the customer's persona and preferences is key. There are powerful tools such as Google Analytics that can help companies measure customer engagement metrics on the website to identify specific areas on the site that may require improvement. For example, using such tools, companies can measure the average time spent by a customer on a web page; the bounce rate (how many visitors came to the site but then quickly left); the average duration of a customer session; the number of pages per session; clickthrough rates on search results (number of times the web site link is clicked as a percentage of the number of times it is shown as part of search results); and top exit pages (the most frequent pages on the site from which the customers exit the site)[14].

---

14. Abbamonte, Kiera. "The 14 Website Engagement Metrics Every Marketing Team Should Be Tracking | Databox Blog." *Databox*, 4 Mar. 2021, databox.com/website-engagement-metrics.

C. **Increasing conversion rates:** Once the customer prospect has crossed the first few steps of the purchase funnel or the sales funnel or the customer journey (awareness; interest; desire as per the AIDA model) and is now taking action to make the purchase, the prospect typically creates a shopping cart on the eCommerce web site. This cart is similar to a basket of products that a shopper uses in a physical retail store. Now the goal becomes that of conversion: To have the customer complete the checkout process and complete the purchase. This last step is more difficult than it seems. As of August 2017, 77.3% of all online retail orders such as shopping carts were abandoned before a purchase was made[15]. By another estimate, the average rate of shopping cart abandonment by customers is as high as 81%[16]. Driving even small reductions in the shopping cart abandonment rate can lead to a direct uplift in revenues. Shopping cart success takes contribution from three areas: the percentage of customers who added items to the cart; the percentage who reached checkout; and the percentage who completed the purchase[17]. It turns out that establishing trust with the customer, giving them a sense of security and comfort when information is sought from them, offering trusted payment gateways, making the navigation between the cart and the store easy, offering guest checkout options are some of the things that can help reduce the rates of shopping cart abandonment[18].

15. Walker, Tommy. "How to Reduce Shopping Cart Abandonment by Optimizing the Checkout." Shopify Plus, www.shopify.com/enterprise/44272899-how-to-reduce-shopping-cart-abandonment-by-optimizing-the-checkout. Accessed 13 May 2021.
16. Shewan, Dan. "13 Ways to Combat Shopping Cart Abandonment." *WordStream*, 27 Aug. 2019, www.wordstream.com/blog/ws/2016/03/17/shopping-cart-abandonment.
17. Walker, Tommy, ibid.
18. Shewan, Dan, ibid.

Back-end support for eCommerce

All that diligence and investment to generate traffic to the eCommerce site, improving content on the site to sustain customer interest and attention, developing means to reduce shopping cart abandonment is not going to mean much if the customer-generated orders cannot be fulfilled and delivered to the customer's satisfaction. Even worse would be situations where the prices shown to the customer do not get honored, or the product items that showed as available when the customer was making the purchase, are no longer available and are stocked out. This requires a tight (systemic) integration between the eCommerce site and the back-end data stores for inventory, pricing, and order management. Specifically, an accessible view of inventory must exist across all locations, such as the warehouse, the distribution centers, and the retail outlets. The company has to have a distribution ability to manage procurement, inventory, orders, fulfillment, warehousing, logistics, and returns. This typically requires a tight integration between the eCommerce front end and the back-end ERP system. Views of customer orders must exist that cover all the customer touchpoints, and the customers must have the ability to monitor shipping and delivery status.

Reducing Costs for the Finance Function

The thrust of serving value from digital in the Finance function comes from the use of automation, or specifically, the Robotic Process Automation (RPA). Recall, RPA bots are software programs that can be designed to automatically execute repeated steps, at scale. Typically, mundane processes that use structured steps and data, and that have stabilized over a period of time so that they have a lower propensity for error in the individual steps, are good candidates for modeling through RPA. Once modeled, the resulting bots (output of an RPA design exercise) can format data into forms that humans or systems can understand; update systems; send out reports and help orchestrate workflows.

The finance function is typically a good department to use for automation through RPA. According to research from the McKinsey Global Institute, 40 percent of finance activities—such as revenue management, general accounting and operations, and cash disbursement—can be fully automated, and another 17 percent can be mostly automated[19].

Here are a few examples of such automation.

Automating the management of Accounts Receivables

Using RPA in the management of Accounts Receivables is typically a good starting point for automation in most finance departments since the supporting process does not have a huge reliance on external documents[20].

19. Chandra, Kapil, et al. "Memo to the CFO: Get in Front of Digital Finance—or Get Left Back." McKinsey & Company, 6 July 2018, www.mckinsey.com/business-functions/strategy-and-corporate-finance/our-insights/memo-to-the-cfo-get-in-front-of-digital-finance-or-get-left-back.
20. Team, ElectroNeek. "RPA for Finance & Accounting: 10 Best Use Cases." ElectroNeek, 7 May 2021, electroneek.com/blog/business-insights/rpa-for-finance-accounting-10-best-use-cases.

Through automation of invoice distribution, monitoring of customer credit profiles, and follow-ups, a company can keep its Days Sales Outstanding (DSO; time taken to collect payments) at a low level, and generally manage its Accounts Receivables well—a requirement to generate good cash flows[21].

Automating the management of Accounts Payable

Traditional processing of Accounts Payable is a manually intensive task[22]. RPA bots can reduce the burden of manual processing by automating activities such as the matching of purchasing orders with the invoice (flagging exceptions for review), the processing of vendor invoices, the payments to vendors, and the provisioning of responses to vendor inquiries[23].

Automating daily P&L preparation

Trading companies track their P&L and risk exposures daily[24]. One RPA installation to automate daily P&L preparation reduced the handling time from 60 minutes to 20 minutes[25].

Automating Financial Planning and Forecasting

The RPA bots can automatically process historical data, collect financial data from multiple departments, and use these to generate financial forecasts and variance reports[26].

---

21. Ibid.
22. Ibid.
23. Ibid.
24. Dilmegani, Cem. "Top 67 RPA Usecases / Applications / Examples [2021]." AIMultiple, 18 Apr. 2021, research.aimultiple.com/robotic-process-automation-use-cases.
25. Ibid.
26. Team, ElectroNeek. "RPA for Finance & Accounting: 10 Best Use Cases.", ibid.

There are other uses cases. For instance, RPA can automate tasks in supplier or customer onboarding; compare supplier prices while preparing a customer quote; identify the best pricing from among a bunch of suppliers; support financial close and reporting by automatically closing out sub-ledgers and creating and sending financial filings to regulatory bodies and individuals[27].

In addition to automation of processes from RPA, the digital is also impacting how finance produces data and reports for organization-wide consumption. Increasingly, finance is enabling consumers of its reports to help themselves through self-service tools: Finance governs and produces the underlying data, ensuring its integrity, and then consumers from sales, marketing, and supply chain planning, for instance, can use this data to generate the reports they need—on-demand[28]. Similarly, the reports that finance generates now benefit from the use of data visualization technology that allows the consuming audience to drill up or down through simple clicks, using live data[29].

With such a preponderance of use cases, the deployment of digital for finance "remains an emerging concept in many organizations"[30].

As the use of automation picks up—and stabilizes—in finance, it will free up the human resource time that now can focus on other value-adding tasks. The use of RPA, for the most part, is an operational efficiency play. As finance organizations get comfortable with a critical mass of processes that are working well through automation, it is likely that the next—digital—frontier for finance will be to now expand the value horizon by using data to deploy artificial intelligence models on the predictive side. Providing more accurate forecasts for company revenues, business unit performance, variance, and consumer demand are all great examples for the next wave.

---

27. Samo, Art. "12 Innovative Use Cases for RPA in Finance & Accounting." KOFAX, 30 July 2018, www.kofax.com/learn/blog/12-innovative-use-cases-rpa-finance-accounting.
28. Chandra, Kapil, et al., ibid.
29. Ibid.
30. Ibid.

Maximizing Benefits from HR

The value footprint from digital in the Human Resources (HR) function hinges around one aspect: Talent.

At the onset, the development of all the digital capabilities to produce value, of course, requires people, and thus the idea that HR has to focus on talent may not come as a big surprise.

But the provisioning of talent for digital has other peculiar aspects that cannot be ignored.

To see this, notice how our Digital Capability Model requires a myriad of skill sets such as for cloud architecture, cloud-based development, Big Data analytics, Artificial Intelligence, integration architectures such as for ESB, iPaaS, and so on. Of course, no single individual is expected to know all the components of the model, but increasingly, the individual has to be versed or at least literate in most if not all the major components. In addition, the use of digital organization models for product engineering that fuse the modern paradigms of design thinking with cloud-native development, Agile, and DevOps puts an additional burden on the requirements for talent. Regardless of the individual requirement niche that a specific talent would fulfill, such as for cloud, artificial intelligence, or cybersecurity, it is very likely that the individual, in addition, would also need to have the ability to work in a product engineering setting.

It turns out that the graduates of most majors—even Computer Science—from most Universities today, maybe lacking the last-mile skillsets: The graduates are not floor-ready; their education and training need to be augmented with additional training that equips them to be productive as an entry-level talent in corporations[31]. This has created a market for IT up-skilling. At the time of this writing, companies such as Revature, Trilogy Education Services, and others partner with

31. Pham, Thanh. "Analyzing The Software Engineer Shortage." *Forbes*, 13 Apr. 2021, www.forbes.com/sites/forbestechcouncil/2021/04/13/analyzing-the-software-engineer-shortage/?sh=1503bd5c321c.

universities and schools such as the University of Texas, George Mason University, University of Penn, Columbia Engineering, George Washington University, Northwestern, and others to offer training programs in Web Development, Data Analytics, User Experience design, and Product Management to interested students to increase the chances of their employment[32, 33].

This suggests a supply shortfall of the required talent for doing digital work. The Bureau of Labor Statistics projected that in 2020, there are estimated to be 1.4 million computer science-related jobs that are available, with only 400,000 computer science graduates with the skills to apply for those jobs[34].

Through another estimate, only 29.4% of candidates for the position of software architect job positions fully met employers' requirements; this number was 39.6% for DevOps job openings[35]. 63% of senior executives surveyed in a report admitted that the shortage of software developers remains an area of concern[36]. By one estimate, the US alone is at the risk of losing $162B through unrealized output if the country's software developer shortfall continues growing at the same pace[37].

When supply lags demand, that typically shows in higher prices. The shortfall in software development talent is no exception to this rule. In some of the most competitive markets for technology, software engineers command earnings upwards of $150,000[38].

---

32. "Programs." Trilogy Education Services, 4 Mar. 2021, www.trilogyed.com/programs.
33. "Universities." Trilogy Education Services, www.trilogyed.com/universities. Accessed 17 May 2021.
34. "Claim of 1.4 Million Computer Science Jobs with Only 400,000 Computer Science Graduates to Fill Them." Econdataus.Com, econdataus.com/claim400k.htm. Accessed 17 May 2021.
35. Kosenko, Natalia. "The Software Developer Shortage in the US and the Global Tech Talent Shortage in 2021." Daxx Software Development Teams, posted Nov 13, 2020, www.daxx.com/blog/development-trends/software-developer-shortage-us.
36. Pham, Thanh, ibid.
37. Kosenko, Natalia, ibid.
38. Liu, Jennifer. "The US Has Nearly 1 Million Open IT Jobs—Here's How Much It Can Pay off to Switch Industries into Tech." CNBC, 6 Nov. 2019, www.cnbc.com/2019/11/06/how-switching-careers-to-tech-could-solve-the-us-talent-shortage.html.

It turns out that in hiring talent for digital, it is not only getting harder—through the supply shortfall—to get qualified talent, it is also taking longer. According to a recent report, in 2016, it took companies an average of 55 days to fill a tech role; this same number jumped to 66 days in 2019[39]. Not being able to fill these vacancies in time comes with a direct impact on revenues. Per the same report, not being able to fill these vacancies leads to an estimated $680 in revenue loss per vacancy per day[40].

Not being able to procure the right talent in time leads to revenue loss for the company, and over time this continued lag can seriously impair the company's competitive advantages. This creates a mandate for HR in the digital world: Hire the right talent with speed, keep the talent engaged and productive; and, sustain its employment (in the US IT industry, staff turnover is around 13.2%, highest among all industries[41]).

Through the emphasis on the customer and her experience, the digital had shifted the focus on the human over the machine. What the above trends in digital talent signify is the continuation of this phenomenon: Digital impresses upon the HR organization to develop policies, processes, workplaces, tools that attract and retain employees. That is, digital, next to customers, has added one more element of human focus: The Employee.

For employees in the 21st century, the lines between personal and professional lives are blurred; these employees will check their work emails over the weekend and check their social media accounts during work hours[42]. As such, modern employees expect to be treated as customers[43]. Such employees expect self-service tools over the web and mobile—that HR needs to provide—that allow them to be more in control of their destinies. They expect abilities to self-manage their HR data, absences, professional emails, and workload; have near real-time

39. Ibid.
40. Ibid.
41. Kosenko, Natalia, ibid.
42. Verlinden, Neelie. "Back to Basics: What Is Digital HR?" AIHR Digital, 26 May 2021, www.digitalhrtech.com/back-to-basics-what-is-digital-hr.
43. Ibid.

access to feedback and training on any device; and have multi-device access to payroll and benefits information[44]. The same demographic—the Millennials and the Generation Z—that is digitally savvy and that is driving the need for bands to do exceptionally well on how they treat their customers, is serving as the employee base today for most organizations, and their percentage in the workforce mix will only increase in the future. Thus it should not come as a surprise that what this demographic expects as a customer is the same thing it expects when serving as an employee.

HR functions can respond to the talent dynamics of the digital age by using a data and artificial intelligence-driven candidate selection process (that increases the chance of an accurate fulfillment) and automating onboarding programs. HR systems can use bots to automate the aggregation of CVs, arranging interviews, and maintaining notes across the selection process[45]. RPA can also reduce the manual steps required in managing payroll that follow company rules and regulations and employee absences. For instance, a 7-week RPA implementation reduced the manual effort in payroll processing by 25% while increasing the processing speed by 85%[46]. An OCR (Optical Character Recognition; an artificial intelligence application) enabled RPA solution can automatically decipher expense receipts, extract important fields from them, and reduce the manual steps an employee has to take to submit expenses[47]. An RPA chatbot, similar to how it answers customer questions in a customer service setting, can be a one-stop-shop to answer most—if not all—employee questions around benefits, performance cycles, improvement opportunities, vacation policies, expense submissions, and so on[48].

In addition to using artificial intelligence to make better predictions, read document submissions, and RPA to automate back-end tasks so that the HR team can spend more time in analyzing data and

44. Ibid.
45. Cem Dilmegani, ibid.
46. Ibid.
47. Ibid.
48. Ibid.

facilitating organization-wide decision making, modern HR also has to enable a digital workplace environment. Recall that the use of the product or the digital engineering model required the use of modern paradigms such as Agile, and DevOps, that focus heavily on team dynamics and collaboration. In almost all cases, they require teams to be not only physically co-located but also set up in a way that enables strong collaboration and communications. What all that implies is that the physical design of the workspace—how desks are located; where the kitchen, playrooms, huddle rooms are located; how the free space is organized; what colors are used for the walls; where television monitors that, for instance, show the team's work throughputs are located and so on—matters crucially for the workforce performance, and motivation. HR today has to drive the enablement of such digital workplaces.

Value Creation from Digital, by Industry Sectors

In the previous section, we saw how value gets created from digital in the functional areas such as inventory management, sales, marketing, customer service, finance, and Human Resources. Most areas of value creation in functions are operational efficiency plays, even though some cases of direct revenue generation such as in marketing, and sales, and digital commerce are catching up with speed.

Generally, a majority of these functional value creation uses cases can apply to most industries. Even so, industry sectors have their own peculiar value-creating digital use cases.

Asset-intensive industries such as Manufacturing have a preponderance of digital use cases where manufacturing operations are made more cost-effective by, for instance, using digital for predictive maintenance—detecting and remediating product or machine operational issues before they become problems, thereby reducing overall costs of operations. Regulated industries such as the pharmaceutical and financial, create value from digital by scaling the use of automation in their back-end and monotonic operational functions and by using analytics to bring down SG&A costs of functions such as R&D.

Non-asset-intensive industries have the luxury of focusing maniacally on customer experience to drive up revenues. Retail overwhelmingly falls in this category. A close cousin is the Consumer Packaged Goods industry, where the focus on customer experience is catching up—especially with the increasing adoption of the electronic D2C channel—even though the need to drive supply chain and manufacturing efficiencies—through the use of digital—continues to exist.

It may be useful to bring the well-known Michael Porter's framework of industry assessment to see how digital has changed industry dynamics. Digital's specific footprint, of course, is going to differ by industry type. Even so, some broad inferences can be drawn. To see this, let us recall the five forces in Porter's framework—the bargaining power of buyers; strength of barriers to entry or the threat of new entrants; the bargaining power of suppliers; the threat of substitute products or services; rivalry among existing competitors[49].

When it comes to the bargaining power of buyers or customers, digital has increased that by orders of magnitude in almost every industry type. As we have considered earlier in this book, the emergence of technology, new digital-savvy customer segments, the need to reduce channel costs through the electronic D2C channel, and so on—all have handed a powerful market-shaping remote control in the hands of the customers. Customers can change brands almost as easily as they can flip channels on their television remote control.

We have also seen that the barriers to entry are not as strong as they used to be in most industry sectors. Digital, we marveled, is a diffuser of most boundaries. The emergence of customer-centricity, the need to give customers the same experience regardless of the sector, and the need to use the cloud, analytics, and product engineering in every industry sector to bring down costs and embrace new sources of revenue have contributed to removal or erosion of fences among industry sectors. Technology investments consume a majority of investments in almost every sector. There are not many business strategies these days in any sector that do not have strong underpinnings of technology and digital. An exception to this observation perhaps are regulated industries such as pharmaceutical, and banking, that involve strong capital investments, are prone to governmental regulations, and can have long payout cycles on investments.

Regarding the bargaining power of suppliers, digital—perhaps again as a diffuser of boundaries and a force that levels the playing field—instead of creating dynamics that promote a rift between the company and its suppliers, has in some cases increased the collaboration between the company and its suppliers, benefiting both. A

49. "The Five Forces—Institute For Strategy And Competitiveness—Harvard Business School." HBS, www.isc.hbs.edu/strategy/business-strategy/Pages/the-five-forces.aspx. Accessed 21 June 2021.

poster child of this behavior is Retail. In Retail, one of the inherent characteristics of the industry is that there is always a fluctuation in demand, and to mitigate the impacts, retailers have to carry some buffer stock that increases the carrying costs of inventory and is a bleed on margins. Reduce the buffer stock drastically, and a retailer runs the risk of stockout, a revenue losing proposition. As a result, a retailer always has to draw a balance and identify optimum levels of inventory to maintain. In modern times, what digital has done is that through collaboration platforms, a retailer shares its demand and order forecast and inventory level data with the supplier, allowing the supplier to reduce its own costs by controlling its production and workflow levels. The retailer benefits because its lead times and on-time delivery rates improve while the frequency of its stockouts reduces. Both the supplier and the retailer win.

Companies across all sectors that can deploy effective customer and product intelligence-gathering engines—using digital—can reap the benefits of introducing better functioning products that address the existing, new, or emergent customer needs better. These product substitutes or augmentations can help the company create distance from its competition. The emergence of the electronic D2C channel can be a powerful substitute for existing models of reaching out and selling to customers. In the case of the Consumer Packaged Goods industry, whose traditional model was to sell to the end consumer predominately using the retail channel—an intermediary—recent trends show an increasing use of the eCommerce channel. Similarly, the emergence of online retail banking is replacing the traditional brick-and-mortar channel.

As far as the impact of digital on the final force—rivalry among existing competitors—is concerned, digital acts as a powerful lever for a company to drive product and service differentiation through the use of cloud (better ability to scale; shift of cost burden from capital to variable costs), analytics (better ability to sense and address the unmet customer need ahead of the competition), product engineering (better ability to respond to market and external change), IoT (sensors and analytics to develop smart products) and so on. Neglecting digital in modern times, as we have seen earlier in this book, can easily make a company exit competition altogether. Thus, as far as competition is concerned, digital can be a powerful lever for a company to leapfrog its competition. In the same breath, underfunding or neglecting critical digital investments slowly but surely can create a path to the company's bankruptcy. In this sense, digital has increased the Darwinian aspect of Michael Porter's final force: The price of neglecting digital can be devastating, while the rewards for nurturing it well can not only help the company create distance from its competition but also enable the company to root the competition out.

Value from Digital in Retail

The retail value chain starts with a producer or the product manufacturer, goes through a wholesaler that buys in bulk, and then passes on to a retailer who breaks bulk by purchasing and reselling smaller quantities of the products to the end consumer[50].

That is, a retailer is an intermediary. And even though Retail has been able to withstand the impacts from the disintermediating effects of digital (recall that digital tends strongly to sideline or eliminate intermediaries altogether in business models), the industry sector has not shown complete immunity either. For one, producers are disintermediating Retail by bypassing it completely to launch their own D2C channels—both electronic and physical—to sell directly to their consumers. Producers do this to exercise more channel control, directly gather information about their customer's behavior and preferences, and test out a few products as well[51]. For instance, Coach sells its products directly to consumers through the retail stores that it directly owns, bypassing its retail partners such as Bloomingdale's, Macy's, and Nordstrom[52]. As noted earlier in this book, Consumer packaged goods companies have launched their own eCommerce platforms to connect directly with their customers, bypassing retailers. (Some of the recent CPG companies have gone for an exclusive electronic D2C model.) This act of disintermediation is impacting retailers as well, quite a few of whom are stepping out of the role of just an intermediary by manufacturing and selling their own brand of products, a concept known as private labeling. Examples are the Great Value line of products from Walmart that competes with producers as varied as Starbucks for coffee, and Chobani for greek yogurt; or the CVS line of over-the-counter products.

50. Donnellan, John. *Merchandise Buying and Management.* 4th ed., Fairchild Books, 2013, p. 2.
51. Ibid., p. 4.
52. Ibid.

Retail continues to add value by selecting what goods customers get to purchase, purchasing them at a lower price from suppliers, gathering customers at a site—physical store, or a website, or a mobile app—to present selected goods to them in attractive assortments, helping customers with their purchase, and then finally by orchestrating the movement of goods through the supply chain: From the producer to the store[53].

Broadly, there are six areas where digital has changed how Retail adds and manages this value creation. They are: customer experience management; digital operations; channel management; marketing; supply chain management; and assortment management.

We now take a closer look at each of these areas.

Customer Experience Management

In the 1980s, a team of researchers found that when people purchase mundane items like soap, toothpaste, toilet paper, and trash bags, they buy habitually, without involving any deliberate decision making, which makes it hard for marketers to use their typical levers such as coupons and promotions, to incentivize shoppers to change their purchasing behavior[54]. But when these same customers are going through a life event such as marriage, graduation, divorce, or a new job, their shopping habits are vulnerable to change[55].

According to a story that appeared in New York Times, Target, a major US retailer, used such a 'vulnerability point' in a customer's life to attempt to change her shopping behavior[56]. Or, to be more specific, incentivize her to visit a Target store or the company's website to make a purchase.

---

53. Niemeier, Stefan, et al. *Reshaping Retail.* Hoboken, NJ-United States, United States, Wiley, 2013, pp. 30-33.
54. Duhigg, Charles. "How Companies Learn Your Secrets." *The New York Times*, 22 Feb. 2012, www.nytimes.com/2012/02/19/magazine/shopping-habits.html.
55. Ibid.
56. Ibid.

Birth records are generally publicly available. A traditional approach to marketing had been to identify new parents and send them birth-related products' advertisements and coupons—clearly, a reactive approach.

What Target, according to the New York Times story, started to do was take a proactive approach. Target had a baby shower registry that was used to identify how shopping habits changed of the prospective mother as she approached her due date—the information was willingly disclosed by the women on the registry[57]. Using the data in the registry, it was observed that the women on the registry were buying unscented lotions in larger quantities around the beginning of their second trimester; in the first 20 weeks, pregnant women stocked on magnesium, calcium, and zinc supplements[58]. Using such a model, it became possible to identify 25 products that enabled an assignment of a "pregnancy prediction" score, along with an expected due date within a small window[59]. This pregnancy-predicting model was applied to every regular female shopper in Target's national database, according to the New York Times report, and it generated a list of tens of thousands of women who were very probably pregnant, and thus could be marketed to for relevant products and offers[60]. It turns out that, using such a scheme, in one case, the company discovered that a girl was pregnant before her father did[61]. To avoid the public relations nightmare of privacy invasion, the company started mixing advertisements for pregnancy products with non-pregnancy ones. So, for instance, ads for lawnmowers were put next to those for diapers; for wineglasses next to infant clothes to give the impression that the ads were random[62]. According to the New York Times report, from the ad campaign, Target's Mom and Baby sales exploded[63].

57. Ibid.
58. Ibid.
59. Ibid.
60. Ibid.
61. Ibid.
62. Ibid.
63. Ibid.

The above is an example of how monitoring the customer behavior and using it to create a more convenient (the same woman customer in addition to buying pregnancy-related products, can shop for other items as well, from the same grocery store, minimizing the number of stores she has to visit for her needs) experience for customers can lead to top-line growth.

But with the advent of the eCommerce channel, it is also important that customers' experience across the physical store and the digital channels—the web, mobile or social media—is harmonized. In fact, all of the brand's touchpoints with the customer must provide continuity of experience through the entirety of the customer's purchase journey covering moments of awareness, interest, purchase, delivery, post-sale consumption, and product return experience. In addition, a brand must use a variety of customer purchase and operational data, augment it with customer attitudinal and other data—often purchased via third parties—to make personalized offers to its customers and provide them with a personalized experience. Customers these days also appreciate the ability to self-serve themselves for things such as product research, product comparisons, initial quote generation, billing and payment monitoring, and customer service request management. Thus, to improve its customers' experience, a brand must also offer self-service tools to its customers.

Digital Operations: 360-degree Demand, Inventory, Customer, and Order Management

Retail operations typically cover the functions of buying, planning, distribution, and if developing private labels, product development as well[64].

Buying entails exploring product offerings in the market, purchasing goods, pricing them for resale, and composing assortments of products that will appeal to the organization's targeted customer base[65].

64. Donnellan, John, ibid.
65. Ibid.

Supporting buying is the planning function that uses historical sales trends to forecast future sales and inventory levels and uses analytics in a major way[66]. Distribution allocates arriving merchandise shipments to stores based on the store's capacity and corrects stock imbalances[67]. Buying, planning, and distribution have a high corporate footprint. The non-corporate—the store or field level—functions are responsible for presenting merchandise appropriately in the store, maintaining the right store inventory levels, and creating proper merchandise assortments in the store[68].

The goal of the entire retail operations is to facilitate a customer purchase by presenting the right product, to the right customer, at the right time, using the customer's preferred channel of purchase. To support this goal well requires complete—we can call them 360-degree—information views of a few core data entities. These entities are: customer demand; inventory; supply; customer orders; and customer interactions.

The 360-degree aspect implies that the data or information views must be holistic. So, for instance, a 360-degree view of customer demand and orders must cover orders from all channels—physical stores, mobile applications, eCommerce website, Call center, and social media. A 360-degree view of inventory must cover information across all locations: warehouse, distribution centers, stores. A 360-degree view of supply must cover information across the entirety of the supply chain—from producers through supplier warehouses, transportation, and reception at the retailer's warehouse or store. A 360-degree view of customer interactions must include interactions across all channels while enabling the retailer to synthesize a view for a customer along her path to purchase—covering points along the customer journey for awareness, interest, purchase, post-sales support, delivery, and customer service.

These holistic information views are critical for the types of sophisticated analysis that are used by marketing to optimize prices and pro-

66. Ibid.

67. Ibid.

68. Ibid.

motions; for lowering inventory costs; for increasing customer sales; for improving assortment margins—and are increasingly becoming table stakes for the modern retailer.

But perhaps more importantly, these holistic information views of demand, inventory, orders, supply, and customer information enable what is known in the trade as an "any, any, any" capability for a retailer: Provide the retailer's customers with an ability to buy from any channel, fulfill from any channel, and return via any channel[69]. Today, this ability has become a core part of a customer's expectations.

Channel Management: Digital Channel Optimization

As noted earlier in this book, customers increasingly are using the digital channels—web, mobile, social media—to interact with the brand for product and service awareness, research, diligence, purchase, delivery monitoring, and post-purchase support. The tenets for optimizing the eCommerce channel, noted earlier, apply to retailers as well. Automation can be used to process returns, manage inventory, interact with customers to answer their queries—freeing up time for resources to spend on value-adding tasks such as replenishing inventory levels for fast turnover items, assisting with markdowns for slow-moving ones. Artificial intelligence-powered predictive product recommendations on the digital channel can help the customer make the final product choice for purchase. Automated generation of personalized content—that is specific to a customer's demographic, shopping habits, preferences, current purchasing frame of mind ("I may be a sports enthusiast, but today I am shopping for office supplies")—harmonized across all channels of interaction can go a long way in prodding the customer along towards purchase.

69. NetSuite.com. "What Is an OMS (Order Management System)? | NetSuite." NetSuite.Com, www.netsuite.eu/resource/articles/erp/what-is-oms.shtml. Accessed 26 May 2021.

Channel Management: Physical Store Optimization

Once data from a retailer's systems for customer relationship management, RFID tagging, Point of Sale, website clickstreams, product catalog, business intelligence, and social media feeds is collected to create holistic views of customer purchases, product, and service usage, it can be analyzed by computer systems for patterns[70]. This pattern detection—known as data mining—can be augmented with artificial intelligence models for clustering, classification, and association to identify monetizable outcomes[71].

Artificial intelligence-augmented data mining can help the retailer identify the purchase patterns of its customers. With these discoveries, supermarkets can design more lucrative placement of products on their shelves and develop offers and discounts on products such that they lead to a better ROI for marketing investments[72]. Customers' locations can be ethically tracked from their in-store use of the retailer's mobile app, and this location information can be used with the aisle or shelf-proximity information to make real-time, in-store offers on the mobile app.

Marketing promotions and campaigns can be customized for customer segments that use groupings—informed from data mining and algorithmic analysis—based on recency, frequency, and monetary spend[73]. Thus, a customer who spends less frequently but purchases high margin products can be treated differently than a customer who purchases frequently but buys low margin products or has smaller basket amounts per purchase[74].

One of the most famous commercial applications of such data use is the 'Walmart-beer-and-diaper' story. Analyzing its customers' purchase patterns, Walmart, according to the story, discovered that when

70. Software Testing Help. "Data Mining Examples: Most Common Applications of Data Mining 2021." https://www.softwaretestinghelp.com/data-mining-examples/. Accessed 26 May 2021.
71. Ibid.
72. Ibid.
73. Ibid.
74. Ibid.

male customers visited the baby department to purchase diapers, it was very likely that they would also purchase some beers[75]. Armed with this discovery, Walmart relocated beers next to the diapers, significantly increasing the sales of both beers and diapers[76].

As another example, a fast-food restaurant fed information from its restaurant CCTVs to algorithms that then determined the right menus to display to its customers[77]. By observing how long the customer queues were, the menu items changed. For instance, if the line was long, the menu screen offered more fast food options, and if the line was short, the menu screen offered dishes that were higher margin and took a longer time to prepare[78].

For a fashion apparel retailer, the use of data and algorithms can allow it to maximize sales by optimizing the floor plans and shelf placements in its physical stores based on consumer buying behavior.

Marketing: Predicting effects of pricing and promotions on sales

One of the common marketing strategies to drive up sales is to use the levers of reduced prices and promotions. But there are hundreds or even thousands of product categories. And marketing budgets are limited. So, how does one find the most optimum price and promotion combination that will maximize sales?

The use of Big Data and Artificial Intelligence-based modeling is helping retailers such as large grocery chains answer this question with increasing accuracy. Using extensive data on customers, their purchase patterns, preferences, market trends, and even weather patterns (that can affect seasonal or daily customer traffic), the artificial intelligence

75. "Data Mining Explained With 10 Interesting Stories." Octoparse, www.octoparse.com/blog/data-mining-explained-with-10-interesting-stories#. Accessed 26 May 2021.
76. Ibid.
77. Ibid.
78. Ibid.

algorithms help retailers identify the most optimum pricing and promotion strategy, by product category; run it in the market; then learn from it to make the next rollout even more successful. Using such a strategy, retailers can take even competition's reaction to a pricing and promotion rollout into account and answer questions such as: Will increasing the average rebate be more effective than increasing the number of items on a promotion? Did price increase or fewer promotions (or both) on a product category cause a drop in sales?[79]. Using such an analysis, a European retailer discovered that increasing the number of promotions instead of reducing pricing will have the biggest impact on store sales and overall profitability[80]. Another leading retailer discovered that only 20 percent of its offers drove additional sales and customer loyalty; using this insight, it reduced the number of monthly promotions from 750 to 300 items; further, through better customer segmentation and using it to make more tailored offers to its customers, the retailer reduced its promotion expenses by 300 million pounds[81].

Using Digital to Optimize Supply Chain

Using bar codes, or RFIDs, retailers can track a single item such as a chapstick or a can of soda across the supply chain from the time the item leaves the manufacturing floor to the time it is scanned through a point of sale (POS) system at the checkout counter[82]. Artificial intelligence algorithms can then use this data to detect issues before they become problems, enabling the concept of self-healing or self-correcting supply chains. The same algorithms can also drive supply chain efficiencies by product category. Thus service levels, logistical costs, lead, and delivery times can be individually adjusted and managed differently for fresh versus dry food items (grocery retailers); for

79. Niemeier, Stefan, et al., ibid., pp. 138-139.
80. Ibid.
81. Ibid.
82. Ibid., p. 36.

basic versus high fashion clothing (fashion apparel retailers)[83]. Such micro-segmentation of the supply chain helps retailers better manage variability in demand, product obsolescence, and stockout, as well as handling and transportation costs[84]. Using such micro-segmentation of the supply chain, a European retailer reduced its inventory volumes by 20 percent and total supply chain costs by 6 percent, in addition to significantly improving product availability in stores[85].

As also discussed earlier, retailers are now collaborating with their suppliers on demand and order forecasts, as well as inventory level data, to help suppliers reduce their own production costs (through more accurate planning) while allowing the retailers to reduce their own inventory carrying costs and stockouts.

Using Analytics to Optimize Assortment

The success of retail stores depends on how attractive and relevant assortments are of the products that are displayed to the customer for purchase. Assortments per product category can be broad and shallow (for example, many product categories on display; multiple brands displayed for a category, but limited brands per category, or limited product lines per brand), or deep (a comprehensive list of brands in the category; full or extensive product lines for the brands in the category)[86]. Thus, Walmart or Target offer their customers a broad selection of merchandise that includes diverse categories such as sporting goods, office products, hardware products, and so on[87]. On the other hand, Dick's, Staples, and Home Depot carry much deeper assortments relative to Walmart for their product categories[88].

83. Ibid., p. 137.
84. Ibid.
85. Ibid.
86. Donnellan, John, ibid., p. 27.
87. Ibid.
88. Ibid.

Fashion apparel retailers have to select their product assortments carefully for various fashion categories. These assortments need to have a good blend of brands, prices, color, fabric, styles, and sizes. Often, an assortment entails combining high and low margin products. What should be the optimal combination in an assortment?

Answer to this question used to be subjective and based on the experience of the buyers and suppliers who typically carry a much richer perspective on categories, or used some basic rules such as cut the least profitable products out from an assortment without understanding how that will impact the sales[89]. Now with Big Data and artificial intelligence algorithms, retailers can use knowledge of customer preferences, macro fashion trends, and how customers make their purchase decisions to identify which customer needs are unmet by the current assortments, and determine product combinations in assortments that will maximize sales, and improve margins[90]. Using such an advanced analytics scheme, a leading grocery retailer changed its assortment by replacing 10 percent of the assortment's low-margin products with higher margin ones, improving both sales (by 2 percent) and margins (by 1 percent)[91].

89. Niemeier, Stefan, et al., ibid., pp. 139-140.
90. Ibid.
91. Ibid.

Value from Digital in Pharma

Similar to many other industries, the sources of value in the pharmaceutical industry come from the leverage of smart sensors, Big Data analytics powered by machine learning, and process automation. The use of smart sensors can be in a consumer setting such as through a wearable device—a smartwatch is an example—or in an industrial setting, where the sensorship resides in a machine and is part of a machine-to-machine (M2M) communication that then feeds into an analytic setting that creates higher efficiencies in the industrial or manufacturing operations. In this context, it might be useful to call out the emerging nomenclature: when sensors are used in a consumer setting such as a wearable device, or a smart home, the prevalent term is IoT; the use of sensors in an industrial or manufacturing setting to create smart machines or smart factories, is known as the Industrial Internet of Things or IIoT (a phenomenon also known as the Industry 4.0, suggesting a new wave of "industrial revolution").

Digital Embrace in Pharma

Heavily regulated, and using "a considerable amount of legacy facilities with aging infrastructures," at least in the Pharma manufacturing side of the house, the Pharma industry has been relatively slow in embracing the full thrust of digital[92]. In a set of survey results reported in 2019, when Pharma manufacturers were asked if they believed their companies would choose to automate manual yet seemingly effective plant

92. Langhauser, Karen. "Pharma Finds Its Digital Groove." *Pharma Manufacturing*, 19 Aug. 2020, www.pharmamanufacturing.com/articles/2020/pharma-finds-its-digital-groove.

processes, 24 percent simply said no[93]. From 2017 to 2019, the percentage that reported that their companies had actually invested in IIoT related digital transformation decreased from 34 percent to 25 percent[94]. These survey results that seem to be skewed towards Pharma's embrace of IIoT—the use of digital or smart sensors in an industrial or manufacturing setting—may not cover other use cases of digital such as the use of machine learning-based analytics, or even the consumer side of sensor devices. In addition, the advent of the pandemic may have caused a shift in thinking about how much value can come from digital in the Pharma industry.

In an interview reported in August 2020, the Novartis CEO said that his company's—Novartis is a Pharma giant—digital investment areas that had "previously seemed like nice-to-have experimental areas that may transform in five years" had suddenly become things that are fundamental[95].

Industry trends in Pharma may be forcing this shift in thinking.

Pharma, like all healthcare, is moving towards outcome-based treatments for patients. Product or drug performance is increasingly available online, along with detailed pros-and-cons based analyses for when it makes sense to use the drug as well as information about the underlying disease[96]. The modern patient—similar to a customer in Retail—is empowered with information gathered from self-initiated due diligence and research. This patient can have a conversation with both the providers and payors on the most effective option for treatment. Issues with care can quickly snowball into a public relations nightmare, thanks to social media and other channels where patients can provide feedback.

---

93. Langhauser, Karen. "Recharging Pharma's Digital Revolution." *Pharma Manufacturing*, 15 Oct. 2019, www.pharmamanufacturing.com/articles/2019/recharging-the-digital-revolution.
94. Ibid.
95. Langhauser, Karen. "Pharma Finds Its Digital Groove." Ibid.
96. Champagne, David, et al. "The Road to Digital Success in Pharma." McKinsey & Company, 20 Mar. 2019, www.mckinsey.com/industries/pharmaceuticals-and-medical-products/our-insights/the-road-to-digital-success-in-pharma.

It may very well be that an increase in the pace of adoption of digital, across the Pharma value chain, is forced by this empowered consumer—the patient—and what she expects from her treatment as well as what she can do if her expectations are missed.

On the regulatory side, the US Food and Drug Administration, with a reputation as an innovation blocker, has taken steps to overcome this perception[97]. The agency has created an Emerging Technology Program that allows Pharma companies to work with it to resolve potential regulatory issues related to the use of novel technology—prior to filing a regulatory submission[98].

As heartening signs of appetite for and ability to pursue digital, when the pandemic hit, Pfizer was able to roll out remote collaboration tools in two weeks that allowed its employees to train themselves using augmented reality via smart glasses and mobile phones[99].

Modern Patient Expectations

The modern patient expects an omnichannel experience that provides 24 x 7 access to personalized care that feeds and benefits from an ecosystem of digital devices such as the wearable watch; facilitates active monitoring of patient's health and an active connection with the providers, even payors, to improve outcome-based care for the patient[100]. An example is a mobile app for managing type-2 diabetes that is connected to a 24x7 glucose monitoring device and that provides metrics-based guidance to the patient and alerts the provider if thresholds are at risk of getting crossed. Or, a smartphone-based electrocardiogram[101].

NeoCare Solutions, a virtual-care app developed by Aetna, allows a neonatal nurse to provide on-demand coaching to new parents return-

97. Langhauser, Karen. "Pharma Finds Its Digital Groove." Ibid.
98. Ibid.
99. Ibid.
100. Champagne, David, et al., ibid.
101. Ibid.

ing home with the newborn[102]. Such apps can help the payor keep the cost of healthcare services low by minimizing the time at or the number of visits to the hospital while at the same time providing care to the patients in a setting of their comfort.

The pandemic has brought optimism on the use of digital in the pharma supply chain: 60 percent of pharma manufacturers, in a recent survey, believe that the pandemic will create higher interest in digitizing the supply chain[103].

Optimism on use of Artificial Intelligence

The optimism path continues on the powerful frontier of value from digital: the use of Artificial Intelligence. Fifty percent of global health companies, according to recent research, plan to implement and broadly adopt AI by 2025[104]. A full 72 percent of healthcare organizations believe AI will be crucial to their business models for the future[105].

The top pharmaceutical companies such as Roche, Pfizer, AstraZeneca, Merck, Sanofi, GSK, Johnson & Johnson, and Bristol-Meyers Squibb have already collaborated with or acquired AI technologies[106]. For example: MIT, Novartis, and Pfizer partnered to transform the drug design and manufacturing process using Machine Learning (ML); GSK leveraged AI to enhance the drug discovery process for COVID-19; Roche leveraged ML to speed up drug discovery, development, and clinical trials; and Abbott launched an AI-powered coronary imaging platform that can detect the severity of calcium-based blockages and measure vessel diameter to increase decision making precision during coronary stenting procedures[107].

---

102. Ibid.
103. Langhauser, Karen. "Pharma Finds Its Digital Groove." Ibid.
104. "AI in the Pharma Industry: Current Uses, Best Cases, Digital Future." Pharmanewsintel.Com, 30 Apr. 2021, https://pharmanewsintel.com/news/ai-in-the-pharma-industry-current-uses-best-cases-digital-future.
105. Ibid.
106. Ibid.
107. Ibid.

As these examples suggest, the best use cases for using AI in Pharma are in the areas of drug discovery, drug manufacturing, increasing the effectiveness of diagnostics, and optimizing processes for medical treatment[108].

Physiological simulation can accelerate drug development; 3-D tissue modeling and computer simulation can help assess the potential toxicity of an R&D initiative; in late development, wearable devices can stream sensor-data streams from in vivo trials to inform registration filings, providing an early indication of real-world effectiveness[109]. The use of AI in R&D has the potential to bring down drug discovery and development costs. Similar to how an inventory planner in Retail can use algorithms to identify other stores where a slow-moving item in the current store can sell better, AI and ML can help identify repurposing of new drugs: the predictive algorithms can help identify other target diseases where a compound that has failed in clinical trials, has the potential to perform better[110]. Again, similar to how AI and ML can create self-healing and demand adjusting supply chains in Retail, AI-ML can help forecast and mitigate under-demand and over-demand issues in the Pharma supply chain and proactively fix issues in the production line[111].

Regarding the use of AI for increasing the accuracy of diagnoses and improving the effectiveness of treatments, these are the individual functionality-based approvals provided by FDA for AI-based algorithms, by year, and specialty: Genius AI Detection, a software device that identifies potential abnormalities in breast images (2020; Radiology); AI for analysis of ultrasound scans (2020; Radiology); for automated detection of diabetic retinopathy (2020; Ophthalmology); for computer-aided triage of CT images (2020; Radiology and Emergency Medicine); for cardiac monitoring (2020; Cardiology); for analyzing lesions suspicious of cancer (2020; Radiology and Oncology); for calculating concussion-index based on EEG (2019; Neurology); for

108. Ibid.
109. Champagne, David, et al., ibid.
110. "AI in the Pharma Industry: Current Uses, Best Cases, Digital Future." Ibid.
111. Ibid.

diagnosing urinary tract infections (2019; Urology); for monitoring vital signs (2019; Hospital Care); for measuring liver iron concentration (2018; Internal Medicine); for diagnosis and treatment of ADHD (2016; Psychiatry)[112].

Electronic Medical Records (EMR) can provide a full medical history and treatment graph for a patient across providers, locations, and payors (similar to a Customer 360-degree view in Retail that provides a complete set of information for a customer across channels and over time). As the use of a digital-medical ecosystem—that involves the use of sensors such as wearable watches—increases, a more comprehensive record becomes available of patient-provider interactions, adherence to prescribed treatment plans, and of course, the patient's health and overall symptoms. These can be automatically translated to treatment-care success metrics such as reduction of glucose counts, blood pressure, and heart rate levels over time so that they can be—also through automation—tied to reimbursements. Cataloging of such a rich data profile can also lead to the gathering of critically useful information about segments of the population. AI-ML models can scour local or regional EMR population databases to predict the onset of diseases. Pharma companies can mine such data sets to automate the generation of pipeline candidates and improve commercialization results of existing launches[113]. Similar to using AI-ML clustering algorithms to improve customer segmentation in commercial companies such as Retailers and Banks, marketing and sales force in Pharma can use AI-ML generated high quality and granular 'target patient' segmentations to identify ahead of time sufferers from a specific rare disease, and proactively reach out to providers to offer treatments to such patients[114].

Automation—The use of intelligent RPA—can be used in Pharma for onboarding new employees, monitoring launches, approving marketing content, and automating workflows[115]. Connectivity between

112. "The Medical Futurist." Medicalfuturist.Com, https://medicalfuturist.com/fda-approved-ai-based-algorithms/. Accessed 29 May 2021.
113. Champagne, David, et al., ibid.
114. Ibid.
115. Ibid.

the trial-sites and patients as well as automation can improve the patient experience in trials while reducing the management overhead in clinical trials from patient recruitment to result submission[116].

Over time it is likely that the use of sensors, devices, AI-ML algorithms, and automation converges in Pharma to increase the effectiveness of treatment outcomes, as well as the quality of experience of everyone involved from patients, to providers, to salesforce personnel, to patient's families, payors, and employees. These same drivers will likely also continue to increase the collaboration among Pharma, patients, providers, and payors. As the involvement of AI-ML in treatment increases, clearly, that will create new requirements for talent, and a refresh of curriculums that produce pharmacists, R&D personnel, manufacturing technicians, doctors, and nurses—of the future.

116. Ibid.

Value from Digital in Manufacturing

In the manufacturing industry, process control is usually established through the use of setpoints, which are the desired or target values for an essential variable. Deviation of the variable value from its setpoint can be used to regulate the variable and control the underlying process. So, for instance, a distributed control system (DCS) can be connected to sensors and use setpoint control to control the flow of material through a plant[117]. This DCS can send the setpoint required by a process to its controller that can then instruct a valve to operate so that the process stays at the desired set point[118].

The Industrial Internet of Things, or IIoT, is an evolution of DCS[119]. The IIoT, like DCS, is a distributed system in that the system's sensors, and analysis and control mechanisms are spread out (distributed) over physical locations. But it allows for a much higher level of sensor use, intelligence, and automation. So much that it is fueling what is being termed as the Industry 4.0—the next wave of the industrial revolution.

IIoT And Its Core Components

At its root, IIoT is composed of the following components:

(A) **Low-cost sensors:** that are pervasively embedded throughout the manufacturing operations, devices and products. Cost-effectiveness is key—the fact that these sensors are low cost allows them to be used extensively and at scale.

117. Wikipedia contributors. "Distributed Control System." Wikipedia, 6 June 2021, en.wikipedia.org/wiki/Distributed_control_system.
118. Ibid.
119. Wikipedia contributors. "Industrial Internet of Things." Wikipedia, 12 June 2021, en.wikipedia.org/wiki/Industrial_internet_of_things.

(B) **Networks:** Affordable, well-functioning communication networks that can support the massive traffic of sensor data and analytics-generated results with minimal latency.

(C) **Big Data Analytics on Cloud:** The simple fact that the modern information cloud infrastructure provides massive computing power and RAM (memory)—again at very affordable price points—means that the volumes and volumes of data captured by the industrial sensors can be processed for patterns and insights using Big Data Analytics or the Artificial Intelligence—Machine Learning algorithms. The insights gleaned from these AI-ML algorithms using the sensor data can then be automatically translated into control decisions that can then be fed back to the production operations in a timely manner. This way, the performance of the products, production facilities, and manufacturing operations can be optimized on a continual basis. Further, since these are AI-ML models, if they are built well, they will train with time and get better at achieving the desired outcomes. That is, not only can the manufacturing operations and the product performance be continuously optimized, they can also show quantum jumps in improvement over time.

(D) **Edge computing:** The centralized cloud infrastructure is designed to gather sensor, device, and production data from multiple operations, plants, and even global manufacturing locations. Doing so allows a company to harvest economies of scale in the use of computing resources. Not to say that the AI-ML models benefit from the centralization of all production data—these models usually are data hogs in that typically, the more is the data that is fed to them, the better are the chances that the models will perform well. But this can create an issue of latency. In order for the AI-ML model-generated feedback mechanism to be embedded as a control mechanism in the operations on the floor, the latency between when the sensor data is generated and when the control actions become available for execution has to be reduced. The model-generated feedback has to be close to real-time. Edge analytics provides the solution. What happens here is that a data gathering and an analytics engine—the Edge Analytics engine—is placed in close physical proximity to the manufacturing operations or the sensor ecosystem so that a controlling response is generated close to when the sensor data becomes available. This Edge Analytics engine typically has trained models loaded on it. The centralized cloud infrastructure—separate from the Edge—then continues to train the models and improve their accuracy.

Use of Artificial Intelligence in Manufacturing

By embedding sensors into the factory's machines, the Bosch Automotive Diesel Systems plant in Wuxi, China, whose main products are diesel rail systems, exhaust gas treatment components, and gasoline fuel pumps for Chinese on-highway and off-road OEMs, was able to measure and integrate data points related to the machine condition and individual cutting tool information such as the tool length, the tool lifecycle, and reasons for why a tool was changed[120]. The company then developed AI-ML algorithms that used these and other data points to predict equipment failures, enabling the company to schedule maintenance operations well before the failure's incidence as well as analyze the process for bottlenecks[121, 122]. The company reports that using such a sensor-fueled algorithm-driven engine has contributed to more than a 10% increase in output in specific areas[123].

120. Farish, Mike. "Illuminating the Way." *Automotive Manufacturing Solutions*, 2 May 2019, www.automotivemanufacturingsolutions.com/illuminating-the-way/38003.article.
121. Amfg. "Industry 4.0: 7 Real-World Examples of Digital Manufacturing in Action." AMFG, 18 Dec. 2020, amfg.ai/2019/03/28/industry-4-0-7-real-world-examples-of-digital-manufacturing-in-action.
122. Farish, Mike, ibid.
123. Amfg. "Industry 4.0: 7 Real-World Examples of Digital Manufacturing in Action." Ibid.

As another example, customers of Siemens Factory in Amberg, Germany, demand 100% quality for the factory's products[124]. As a result, the Printed Circuit Boards (PCBs) used by the products had to be tested using X-Rays during the manufacturing process to ensure that the solder contact was perfect[125]. The problem with this approach was that since the X-Ray tester required a long throughout time, it became the factory's bottleneck[126]. Buying an additional X-Ray machine would set the company back by half a million Euros; not to say that integrating the new machine into the manufacturing environment would require additional effort[127].

To mitigate the issue, Siemens developed a Big Data algorithm. The company collected solder data such as from the Solder paste Printer, Solder paste Inspection, and the Automatically Optical Inspection, and fed the collected data to its algorithms hosted on its centralized Cloud System Mindsphere[128]. On the cloud, the algorithm was trained. Once the algorithm's predictions were found to be within acceptable limits, the AI-ML model was put on an Edge application that was responsible for the data acquisition from the machine sensors, the data pre-processing, and that provided a prediction for the PCB production. The prediction that this Edge framework provided was a binary answer to the question: "Is additional testing via X-Ray necessary for the board (that has been produced)?" If the answer provided by the AI-ML trained model was "No," the expensive X-Ray testing step was skipped. Using this scheme, the factory reduced its test efforts for the X-Ray machine by up to 30% in the first step[129]. The factory personnel believe that continued training of the AI-ML model can lead to a point where "zero X-Ray tests are needed"[130].

---

124. Staufer, Helmut. "OT Meets IT—Siemens Factory in Amberg, Germany | Ingenuity | Siemens." Ingenuity.Siemens.Com, 20 May 2019, ingenuity.siemens.com/2019/05/ot-meets-it-siemens-factory-in-amberg-germany.
125. Ibid.
126. Ibid.
127. Ibid.
128. Ibid.
129. Ibid.
130. Ibid.

Used in such a fashion, it is leading to the concept of smart factories that can self-optimize their performance; learn from and self-adapt to new conditions in real-time or close to real-time; and run production processes autonomously[131]. According to a Deloitte study, between 2015 and 2018, smart factory initiatives brought an average increase of 10% in production output, an average increase of 11% in factory capacity utilization, and an average increase of 12% in labor productivity[132].

The IIoT and related digital use cases in Manufacturing cover real-time equipment monitoring; in-line quality testing; sensor-based optimization of energy, water, and waste; Big Data analytics fueled visualization for production command centers; automated conveyance for continuous material flow; and use of active or passive sensors tagged to assets, for instance, using RFID, to track assets real-time[133].

Digital Twin

As powerful as above use cases sound, they still do not cover another marvel of digital in Manufacturing: The Digital Twin.

Historically, in a manufacturing production operation, variables such as temperature, material or fluid density, pressure, and other physical characteristics have been fed into operational controls. The information graph used by these operational controls—whether system-based or manual—*has been digital.* So it begs the question, what is the novelty now in what has been introduced by the modern wave of digital—or specifically, through its incarnation, the Digital Twin?

Leaps in modern computing power, AI-ML algorithms, sensor-available data have now made it possible to design, calibrate and

131. "2019 Deloitte-and-MAPI-Smart-Factory-Study." Deloitte, www2.deloitte.com/content/dam/insights/us/articles/6276_2019-Deloitte-and-MAPI-Smart-Factory-Study/DI_2019-Deloitte-and-MAPI-Smart-Factory-Study.pdf. Accessed 7 June 2021.

132. Ibid.

133. Ibid.

test modern products, machines, processes, or even entire production facilities *before they are built or manufactured in the real world*[134]. Today, all of the parameters critical for the functioning of the modern production facilities are sent as signals to the computers and servers, creating a continuous inventory of the production facility's current state at any given point in time[135]. The scope of simulation coverage from these parameters—thanks to the low-cost sensors—is very high. This increase in coverage allows Manufacturing and other companies to use the parameters to not only improve productivity, safety, and alignment to new requirements—something that they have long been doing—but also model the design, manufacture, operation, and usage of the new products[136]. Now, product prototypes can be virtually modeled, and their operational characteristics ascertained *before* any physical prototypes are ever produced. The use of such digital twins of the (physical) prototype increases the speed with which products can be conceived, simulated, and manufactured[137]. Product developers can toy with and compare multiple product configurations using the Digital Twin versions of the products at a rate higher than that afforded by the use of the physical models[138]. The Digital Twin of the product can act like the product's shadow throughout the value chain—from design, through production, operation, servicing, and even recycling—providing the company crucial insights on the product, its performance, and production[139]. Canadian startup Electra Meccanica was able to test and optimize all the elements—from mechanical to electronic, software, and those related to the system performance—of its "Solo" electric car in advance using the digital twins[140].

134. "Digital Twin." Siemens.Com Global Website, new.siemens.com/global/en/company/stories/research-technologies/digitaltwin/digital-twin.html. Accessed 7 June 2021.
135. Ibid.
136. Ibid.
137. Ibid.
138. Ibid.
139. Ibid.
140. Ibid.

The digital twins can be used to design not only products but entire factories. For instance, Siemens uses digital twins for the design of products such as gas turbines as well as its factories[141].

It should come as no surprise then that Gartner predicts that by 2021 half of all major industrial companies will be using digital twins[142]. Another market research conducted by Grand View Research expects the market for digital twins to reach $26 billion by 2025, at an annual growth rate of around 38 percent[143].

141. Ibid.
142. Ibid.
143. Ibid.

Value from Digital in Banking

A few years ago, JPMorgan's CFO proclaimed: "We are a technology company"[144]. At the time, the bank had a team of 40,000 technologists, with a technology budget of $9 billion[145]. A little earlier, Lloyd Blankfein, CEO of Goldman Sachs, was reported to have said that his bank was a "tech company"[146]. At the time, Goldman had 27% of its workforce as engineers and programmers—a number that was greater than the number of tech personnel employed at Facebook; and greater than the entire payrolls of Twitter and LinkedIn[147].

Contributing to this heavy use of technology by the banking industry are a few factors that are impacting the industry. These factors can be categorized into three sets: Channels and Consumers; Competition; and Compliance.

Channels and Consumers in Banking

The channels used by the modern banks include the traditional—bank branches, ATMs, contact centers—and the new—the online banking, and the mobile banking app[148].

However, the digital channels are attracting more customer traffic. According to a 2018 Deloitte study: Bank of America received more

144. Crowe, Portia, and Matt Turner. "JPMORGAN: 'We Are a Technology Company.'" *Yahoo Finance*, 23 Feb. 2016, finance.yahoo.com/news/jpmorgan-were-technology-company-141156910.html.
145. Ibid.
146. Marino, Jonathan. "Goldman Sachs Is a Tech Company." *Business Insider Australia*, 13 Apr. 2015, www.businessinsider.com.au/goldman-sachs-has-more-engineers-than-facebook-2015-4.
147. Ibid.
148. Srinivas, Val, and Angus Ross. "Accelerating Digital Transformation in Banking." *Deloitte Insights*, 9 Oct. 2018, www2.deloitte.com/us/en/insights/industry/financial-services/digital-transformation-in-banking-global-customer-survey.html.

deposits from its mobile channel than from its branches; in all countries, across all generations, the digital channels are being used more frequently than the branches; and 22% of the more than 17,000 surveyed banking consumers were using the online and the mobile app channels more than ten times per month to take care of their banking needs[149].

In the early days of digital banking, Bank of America was able to acquire 1 million new online customers in almost a decade; now, its banking app, Erica, hit the same number within three months of rollout[150].

COVID-19 appears to have strengthened this shift towards digital.

In a 2020 survey conducted by PwC, 27% of the US consumers said that the pandemic had made them more likely to use their bank's website; 23% said the same for their bank's mobile app[151]. Nearly a quarter (24%) said that they are less likely to use the branch office of their bank[152].

While showing a preference for the digital channels, these consumers are also less likely to stick around. According to the Everfi report, 53% of consumers of financial institutions have changed their primary financial provider, and an additional 9% said that they are considering it[153].

As a hypothesis for what is encouraging consumers of financial institutions to switch, it is possible that the banks still have some distance to cover when it comes to developing an emotional connection with their customers. According to Harvard Business Review, emotionally connected customers are 35 percent more valuable than highly satisfied customers, and according to the just mentioned Deloitte study,

149. Ibid.

150. liferay. "7 Reasons Why Financial Services Needs Digital Transformation to Survive." Liferay, 11 Feb. 2021, www.liferay.com/blog/en-au/digital-transformation/7-reasons-why-financial-services-needs-digital-transformation-to-survive.

151. PricewaterhouseCoopers. "Many Consumers Think Their Banks Are Doing a Good Job. In a COVID-19 World, Will That Be Good Enough?" PwC, www.pwc.com/us/en/industries/banking-capital-markets/library/consumer-banking-survey.html. Accessed 8 June 2021.

152. Ibid.

153. Dilmegani, Cem. "Digital Transformation for Banking in 2021: In-Depth Guide." AIMultiple, 1 Jan. 2021, research.aimultiple.com/banking-dx.

banks lag behind other brands in building such emotional connections with their consumers[154].

The modern customer—as we have outlined in the earlier part of this book—being a digitally savvy user, has some core set of expectations from a brand. It is likely that these expectations—such as expecting an emotional connection with the brand—for a major part are formed by the interactions the consumer has with the digital giants: Apple, Google, Facebook, and Amazon.

And it just so happens that each of these giants is making or has made a play in banking.

Competition in Banking

If banks have become tech companies, the tech companies are making a foray into banking. Cautiously.

Per a 2020 CNBC report, Google plans to introduce consumer bank accounts in collaboration with Citibank and a California-based credit union[155]. In 2019, Apple debuted a credit card—in partnership with Godman Sachs—that it integrated with its digital wallet[156]. Facebook wants to introduce a digital currency that increases the speed of global payments, while reducing their costs[157]. Facebook aspires to consolidate its payment products under a new brand called Facebook Pay[158].

Per the CNBC report, Amazon was reportedly in talks with banks like J. P. Morgan to launch its own checking accounts[159]. Amazon had

154. Srinivas, Val, and Angus Ross. "Accelerating Digital Transformation in Banking." Ibid.
155. Browne, Ryan. "Big Tech Will Push Deeper into Finance This Year—but Avoid the 'headache' of Being a Bank." CNBC, 3 Jan. 2020, www.cnbc.com/2020/01/03/big-tech-will-push-into-finance-in-2020-while-avoiding-bank-regulation.html.
156. Ibid.
157. Ibid.
158. Ibid.
159. Ibid.

earlier set up a student loan scheme with Wells Fargo in 2016, even though later, Amazon shuttered the scheme[160].

What all this suggests is that the barriers to entry in banking are perhaps more fallible today. As digital becomes the driving force—as suggested by the massive technology footprints of the banks—for revenue growth, and even cost reduction, and as consumer experience becomes the biggest opportunity—and threat—the banks are challenged to provide to the consumer what the consumer expects from other non-banking brands. And in doing so, the industry specificity—the banking vertical-specific context and knowhow—matters less and less.

It is interesting to note, though, that none of the Big Tech players are looking to become banks themselves. They are pursuing their forays into banking through partnerships with other banks. A possible reason for doing so could be that getting and maintaining a banking license comes with its own overhead of compliance and regulatory approval. Further, the return on capital for a tech company is higher than that for a lender[161]. For what it is worth, the onus of compliance and regulatory requirements in banking may be serving as the last barrier to entry for the industry.

Compliance Requirements for Banking

The financial crisis of 2008 brought on higher scrutiny of banks and their operations. Banks had to hire more staff to keep up with the new laws and regulations that resulted[162].

By one estimate, between 2008 and 2018, banks spent more than $321 billion on enforcement actions, fines, and settlements[163]. Banks' expenditure on compliance can run as high as $270 billion per year[164].

---

160. Ibid.
161. Ibid.
162. internationalbanker. "The Cost of Compliance." *International Banker*, 7 Nov. 2018, internationalbanker.com/technology/the-cost-of-compliance.
163. Ibid.
164. Ibid.

For most banks, 10 percent or more of their operating costs can be attributed to compliance; by some estimates, the regulatory costs could double between 2018 and 2022[165].

In summary, the banking industry faces the challenges of increasing adoption and growth of the new digital channels; meeting the aggressive demands and expectations of the emergent customer segment that wants more and more from their use of digital; competition from the tech giants who know and understand digital much better than the banks do; and the increased load of compliance.

Now, let us see what levers the banks are using to respond to these challenges.

Artificial Intelligence and Automation in Banking

AI and automation are increasingly going hand in hand. AI typically benefits from volumes of data, and automation can help with the automatic ingestion of data and its translation to desired data formats that the AI models can then readily consume. Now let us see how AI is creating specific value in banking and finance.

AI for Trading

Kensho, a machine learning platform founded in May 2013 by Nadler, a Harvard-trained Ph.D., crawls through troves of data about markets searching for correlations between world events and the impacts these events have on asset prices[166]. The platform's data repository—one of the world's largest—can be searched intuitively, and the results viewed

165. Ibid.

166. Gara, Antoine. "Kensho's AI For Investors Just Got Valued At Over $500 Million In Funding Round From Wall Street." *Forbes*, 15 Dec. 2020, www.forbes.com/sites/antoinegara/2017/02/28/kensho-sp-500-million-valuation-jpmorgan-morgan-stanley/?sh=4ee4d9635cbf.

through data visualization[167]. Questions can be asked of the platform in plain English, such as: "Which cement stocks go up the most when a Category 3 hurricane hits Florida?" "When Apple releases a new iPad, which Apple supplier's share price goes up the most?"[168]. These types of questions required several analysts many days; Kensho can execute the task in a matter of minutes[169]. Using the Kensho platform, traders quickly learned that populist votes such as the Brexit historically have led to an extended drop in the local currency; a data point that actually checked out: The pound actually plunged after Brexit[170]. Kensho's machine learning and data analytics platform are used by the leading financial institutions like J. P. Morgan, Morgan Stanley, and Bank of America; S&P Global announced a deal in March 2018 to acquire Kensho for about $550 million[171].

AI for Underwriting and Credit Decisions

Credit history is a valuable asset for a consumer looking to apply for a credit card. The credit profile carries even more value when one considers that according to a recent study, 77% of consumers prefer the use of a credit or debit card to make payments while only 12% preferred cash for doing so[172]. But not everyone may have a track record strong enough to create a decent credit history profile. Millennials fall in this category, as also some other underserved market segments. Tapping into these underserved market segments can increase revenues for the lending companies such as banks. But of course, this has to be done with caution, as these segments—because they do not have

167. Ibid.
168. Ibid.
169. Ibid.
170. Ibid.
171. Schroer, Alyssa. "AI and the Bottom Line: 15 Examples of Artificial Intelligence in Finance." *Built In*, 28 May 2021, builtin.com/artificial-intelligence/ai-finance-banking-applications-companies.
172. Ibid.

a traditional set of information on their credit history—present a risk of default to the bank.

Artificial Intelligence is helping banks and lenders tap into this underserved market. ZestFinance has developed an automated machine learning platform that uses thousands of data points to help banks in their credit decision-making process[173]. According to ZestFinance, using the machine learning-based underwriting platform, lenders for auto loans have cut their losses by 23% annually[174].

Underwriter.ai assesses the credit risk of applicants for consumer and small business loans by analyzing thousands of data points from credit bureau sources; the company's platform reduces the risk of default by applying machine learning to find patterns and categorize loan applications as good or bad[175].

AI for Personalized Banking

In general, the common use cases of using AI to provide personalized experiences to customers such as using AI for superior and granular customer segmentation, to predict lifetime scores and using them to guide marketing campaigns, providing product recommendations on the online channel, and so on, apply to the banking and the financial industry as well.

Similarly, the use of intelligent automation—combination of AI with RPA—to reduce call center costs applies to the banking and financial industry as it does to any other industry that uses call centers to service customer requests. Typically, the play here is to empower the customers to help themselves by giving them self-service options that then can reduce the volume of transactions handled by the call center (because customers call less). The unit costs per transaction can be reduced by reducing the number of things the more expensive cus-

173. Ibid.
174. Ibid.
175. Ibid.

tomer service representatives have to do and by delegating the front office tasks to intelligent—and automated—workflows.

The chatbots can be more effective in servicing the end customer if these bots know better what the customer is looking for and have increasing abilities to tap into the organizational sources of information to service the request. In general, the effectiveness of a customer serving channel such as a bot can be increased by strengthening five criteria: Instantaneousness of response; the intelligence of response; relevance of response; ease of access to the channel such as being able to use the channel through voice requests; access to the channel-enabled service over multiple digital platforms such as online, mobile, and social media.

AI can help the success of automation by seeding intelligence in it.

KAI is a conversational AI platform developed by a company called Kasisto that powers omnichannel bots and virtual assistants with deep domain expertise across web, messaging platforms, mobile, wearables, and IoT devices[176]. These AI-backed bots can engage with customers in contextualized and personalized conversations to fulfill their requests as well as predict their needs[177]. In 2017, TD Bank announced that it would integrate Kasisto's conversational AI platform into the bank's mobile app[178].

Using the AI-powered interactive chat interface, the TD mobile app users can review their transaction histories, monitor spending levels, check account information, and get instant answers about questions such as: How much did I spend on a recent getaway? What was my largest transaction last week? How much did I spend on groceries or coffee last month?

AI models learn with experience—creating a potentially unlimited upside for the technology. So, say, a model has been trained on a bunch of customer profile data points to predict whether a customer will select a product X or not. Over the course of usage, it so happens that the model is right only seven out of ten times; that is, seven out of ten

---

176. Kasisto. "TD Announces Conversational AI Agreement With." Kasisto, 30 Apr. 2021, kasisto.com/press-releases/td-announces-conversational-ai-platform-agreement-with-kasisto.
177. Ibid.
178. Ibid.

times the customer actually selected the model recommended product. The model now will learn from this experience to reduce the 30% non-success rate. This improvement comes with relatively little human involvement. What this does is create the potential of fully automating call centers one day.

AI for Fraud Detection and Cybersecurity

Banks and financial service providers store sensitive data about customers and their financial transactions, making them an attractive target for cyber attacks. By one estimate, every day, major financial institutions face hundreds of thousands of online attacks, with multiple incursions each second[179]. As it is heavily regulated, financial institutions need to comply with more than 800 cybersecurity laws and standards, making management of cybersecurity a ripe case for—intelligent—automation[180]. Only a few years ago, Equifax, the American consumer credit reporting agency, agreed to pay up to $700 million in damages in response to user data stolen from its systems[181].

The company Vectra has developed an AI-powered cyber-threat detection system that automates threat detection, identifies hidden attackers specifically targeting financial institutions, helps speed up post-incident investigations, and identifies information that has been compromised[182]. The company, using this system, was able to prevent malware attacks on a prominent securities exchange[183]. Another company, Shape Security, uses AI-enabled bots to help banks detect compromised login credentials, instantly alerting both companies and customers of security breaches[184].

---

179. Rice, Mae. "Cybersecurity in Banking and Finance." *Built In*, 27 Feb. 2021, builtin.com/cybersecurity/cybersecurity-banking-financial-services.
180. Ibid.
181. Ibid.
182. Schroer, Alyssa, ibid.
183. Ibid.
184. Ibid.

Use of Automation and AI for Regulatory Compliance—Reg Tech

We discussed earlier how the regulatory requirements on banks and financial service providers have become more onerous, especially since the financial crisis of 2008.

RPA and artificial intelligence, in combination, are helping banks meet their regulatory requirement burden in a cost-effective and less error-prone way. The technology stack that has emerged is known as Reg Tech, short form for Regulatory Technology. While not one single thing, Reg Tech has the following capabilities:

Automation—using RPA—of the process of collecting and mining of data and transforming the data to a form that is conducive to the generation of management reports. Automation of manual reporting and compliance processes such as for Know Your Customer requirements[185]. Routine compliance monitoring tasks can be automated, as also identification of critical variances—freeing up the time of the audit staff for analysis instead of manual collection and reconciliation of data[186].

Visual analytics tools that source firm data and present it in an easy-to-understand format for management[187].

Big Data analytics tools that can mine existing data sets, for example, documents as well as structured data in tables, to, for instance, proactively identify compliance red flags or to ease the execution burden of the current set of compliance activities.

185. Wikipedia contributors. "Regulatory Technology." Wikipedia, 5 May 2021, en.wikipedia.org/wiki/Regulatory_technology.

186. ey-Regulatory-Technology-Regtech.pdf. EY.COM, assets.ey.com/content/dam/ey-sites/ey-com/en_us/topics/financial-services/ey-regulatory-technology-regtech.pdf?download. Downloaded 11 June 2021.

187. Ibid.

Use of IoT in Banking

Using the device data generated by ATM kiosks, banks can determine usage volumes and use it to increase or decrease the installation of ATMs in a given region[188]. Banks can adjust financial terms in their loans to the agricultural sector by using IoT data to analyze agricultural output, crop yields, and crop performance, potentially strengthening their relationship with the farming customer[189].

Telemetry data provided by the credit-card swiping location can alert the bank about an unwarranted location—information that the bank can use to reject a fraudulent transaction[190]. Sensor devices installed in the borrowers' warehouses can help the lending bank monitor inventory levels and sales, helping the bank ensure that its loans get repaid and also observe any deviant behavior on the borrowers' part[191].

Internet IoT, the first-ever Internet of Things bank platform, enables people to connect smart devices to their bank account[192]. These devices automatically respond to a user's bank balance[193]. For instance, if a user goes over his spending limit, one such wearable device delivers an electronic shock to him on his wrist[194]. Another device will automatically turn down the user's home's thermostat to a user-defined temperature if the user's account's balance drops below a self-imposed limit[195]. Such devices—in helping a bank's customers navigate their lives better—can deepen the customers' relationship with the bank.

188. Ceo. "The Use of IoT in Banking Industry." *The CEO Views*, 30 June 2020, theceoviews.com/iot-in-banking-industry.
189. Ibid.
190. Ibid.
191. Ibid.
192. Malik, Muhammad. "World's First Internet of Things Bank Gives You an Electric Shock If You Spend Too Much Cash." *Information Security Buzz*, 19 May 2016, informationsecuritybuzz.com/articles/worlds-first-internet-things-bank-gives-electric-shock-spend-much-cash.
193. Ibid.
194. Ibid.
195. Ibid.

Biometric sensors can be used to gather user data when the user is entering a bank and provide this information to the primary system, potentially reducing the amount of time the user has to spend in the bank to complete the transaction[196]. Biometrics data can also be used to accelerate the credit underwriting process[197].

Within their organizations, banks can use IoT devices to serve their employees better by tracking the employees' movements and work habits and using this information to inform better work policies[198]. Banks can also use IoT data to reduce building management and utility costs[199].

Blockchain and Bitcoins

A typical database stores information in rows and columns. Blockchain, another type of database, stores information in blocks that are chained together. New data is entered into a fresh block. Each block is given an identifier—its own hash code generated by a mathematical function—and stores the hash code of the block preceding it[200]. If the information in the block is changed, the block's hash code changes as well. The chain of blocks can be stored in a distributed manner—over a network of computers—called nodes—dispersed across different locations. Each block has a storage capacity that, when filled, causes a new block to be added and chained to the previously filled block, creating a chronology of data termed the blockchain.

The entirety of the data in the blockchain since its inception is stored in each node. This way, if one node has an error, it can use other nodes as a reference to correct itself, and no single node can alter infor-

196. Ceo. "The Use of IoT in Banking Industry." Ibid.
197. "Digital Transformation in Financial Services." BDO, Apr. 2019, www.bdo.com/insights/industries/financial-services/digital-transformation-in-financial-services.
198. Ibid.
199. Ibid.
200. "Blockchain Explained." Investopedia, 1 June 2021, www.investopedia.com/terms/b/blockchain.asp.

mation captured within the network[201]. This provides a mechanism to capture transaction history in an irreversible and transparent way. Further, since the transactions or pieces of information are captured across a network of computers, distributed geographically, there is no single entity that has control over the information, and the information or transaction history is visible to all.

Blockchains are referred to as distributed ledger technology since the transactions or the information is not only recorded as in a ledger but is also distributed across nodes.

Consider the technology's resilience against a hack. Say, a hacker, in her attempt to alter the blockchain, alters her own copy. Other nodes, when cross-referencing with each other to validate information, would see the hacker's copy stand out and mark the standout version as illegitimate. To overcome this, the hacker would need to alter 51% of the copies in the blockchain so that her copy becomes the majority one and thus the agreed-upon chain—an effort prohibited by the volume of computing resources that would be required to accomplish this[202].

Blockchain as a technology was first outlined in 1991 by two researchers, Stuart Haber and W. Scott Stornetta, who worked on developing a cryptology-based secured chain of blocks that would prevent tampering with the timestamps of documents[203]. The researchers upgraded their system in 1992 to accommodate more documents in a single block[204]. In 2009, Satoshi Nakamoto released a white paper on Bitcoin, the first application of the digital ledger technology, and that aimed to increase digital trust leveraging the decentralization aspect of the technology and the fact that no single entity—but only a bunch of computers—seems to be in control[205].

Once a transaction is encountered, thousands of computers rush to validate the transaction, and the transaction is recorded post val-

201. Ibid.
202. Ibid.
203. Iredale, Gwyneth. "History of Blockchain Technology: A Detailed Guide." 101 Blockchains, 21 Dec. 2020, 101blockchains.com/history-of-blockchain-timeline.
204. Ibid.
205. Ibid.

idation[206]. The distributed aspect of this computer network provides a 24x7 "ready to accept" transaction feature to the blockchain technology. This can be a valuable capability to augment a bank's support hours. For example, using blockchain technology, potentially, checks could be deposited any time, over the weekend and off-hours. Further, while a bank may take days to complete a deposit transaction—especially for international transactions—using blockchain technology, the transaction can be completed in minutes[207].

But of course, there are issues with the technology that impact its broad adoption.

The validation of the transactions is an expensive process. It can consume significant computing power, and as a result, build up electricity costs. For example, such validation for bitcoin transactions—known as the "proof of work"—cost about $5,000-$7,000, excluding the computer hardware expenses, per coin[208]. People who have such computing networks—known as miners—and thus are willing to incur these expenses are rewarded with a bitcoin when they are able to add a block to the bitcoin blockchain[209]. Currently, there are 687,071 bitcoin blocks out there, and a bitcoin is worth about $37,132[210]. The miners' reward for successfully completing one block is halved every 210,000 blocks—an average of every four years[211]. (The pay-offs from bitcoin mining have to consider the cost of electricity, and hardware, against the potential revenue upside. By one estimate, in 2020, a modern Bitcoin mining machine, costing a few thousand dollars, can earn about $8 in Bitcoin revenue every day[212].)

---

206. "Blockchain Explained." Investopedia, ibid.
207. Ibid.
208. Ibid.
209. Ibid.
210. Google Search for "how many blocks are there in bitcoin". www.gogle.com. Accessed 11 June 2021.
211. "What Happens to Bitcoin After All 21 Million Are Mined?" Investopedia, www.investopedia.com/tech/what-happens-bitcoin-after-21-million-mined. Accessed 11 June 2021.
212. Tuwiner, Malcolm Cannon Jordan. "7 Reasons Bitcoin Mining Is Profitable and Worth It (2021)." www.buybitcoinworldwide.com, 10 May 2021, www.buybitcoinworldwide.com/mining/profitability.

The onus of this required validation impacts the speed with which the transactions can be processed. By one estimate, the bitcoin's network takes about ten minutes to add a new block to the blockchain, and that the blockchain network can manage only about seven transactions per second[213]. By contrast, Visa can process about 24,000 transactions per second[214].

Perhaps sources of energy such as solar will help bring down the costs of using the technology and help increase its adoption at scale.

---

213. "Blockchain Explained." Investopedia, ibid.
214. Ibid.

Digital, Looking Ahead

Digital has emerged as a powerful force in modern times. It is changing how humans behave as consumers, how companies need to define their product and service offerings and take them to market. This book has highlighted the reasons why digital has become the force it has and the specific levers companies can use—are using—to tap into the opportunities created by digital.

Here are some closing thoughts on where digital might be taking all of us.

Thought number one: Digital will continue to act as a force of empowerment. Even though empowering both the consumer and the brand, it likely will show a bias towards the consumer versus the brand.

Digital empowers both the human and the company the human works for. The human is empowered as a consumer through the powerful apps in his hands, on his powerful devices that are connected on fast networks. The human is empowered through having a powerful voice on social media where he can promote or bring down a product or a brand. The human is empowered through having an unprecedented ability to conduct information research over the increasingly accurate internet search engines. The human has 24 x 7 access to the brand, to shopping, and to conduct almost any transaction. The human is empowered because he does not have to use an intermediary to conduct his transactions and is now more in control of his interactions with the brand. As evidence of this empowerment, healthcare is converging towards patient outcomes; retail and most of the consumer-serving industry sectors are having to continuously identify and respond to the next unmet customer need and to systemically reduce experiential gaps in their customers' journeys.

There is another aspect of human empowerment through digital. It has to do with providing access and ability to the marginalized sections of the population. Through Blockchain and bitcoins, those sections of the world population, where the banking structure is weak, or banks do not exist at all, but people have connectivity to the internet, can conduct financial transactions with each other, and the rest of the world. Through digital, these marginalized sections can be brought into the world economic mainstream, creating seeds for development. Digital through mobile apps is already helping the impoverished farmers in India get subsidy payments directly from the government—eliminating carvings from the intermediate brokers—and to sell their produce in the marketplace at beneficial prices.

The organization—and the brand—have been empowered through digital by getting the means to be able to do more with less—such as through the use of Cloud that allows companies to avoid making huge capital investments upfront to respond to elasticity in user demand; through the use of RPA to automate backend processes and reduce the cost of operations; through the use of automated-intelligence that allows the companies to circumvent expensive research and development and use machines instead to identify the next differentiating product or drug molecule.

At times, customer empowerment will overshoot. At such times, the customers may wield higher than usual influence over the brand or its products. At other times, the company empowerment will overshoot, such as when a company reduces its human workforce because, for instance, the machines can do the same job more efficiently and at much lower costs. In the final scheme of things, the empowerment of the customer will likely hold more sway than that of the brand.

Thought number two: Digital will continue to level the playing field for all involved. This includes industry sectors, organizational departments, and even its technology footprint.

Digital has shown an almost consistent tendency to shatter boundaries and level the playing field. In this book, we saw the silo-breaking aspect of digital on a few fronts.

First, we saw how customer-centricity and empowered consumers are reducing the strength of the barriers to entry that a vertical may have historically enjoyed. The consistent experience that the modern consumer expects from a brand across different industry sectors is driving homogeneity in how all types of brands—need to—treat their customers. As a result, the silo-protecting factors for a vertical are perhaps less relevant today in a digital economy for the simple reason that it is the same digitally empowered consumer that is defining how brands—traversing all sectors—should behave. Second, digital has also become a force for removing silos that exist within the organizations. Modern organizations are starting to coalesce around the need of the customer. We saw how cross-function collaboration—transcending departmental boundaries—such as that created in a product engineering team is becoming the need of the hour and paramount for product success. It is now required of brands to use the best organizational—and cross-functional—talent to deal with the ambiguous and perennially uncertain customer expectations and market and technology trends. As evidence, the Voice of the Customer, or VOC, is emerging in companies as a unit that breaks down organization silos to focus relentlessly and in a cross-functional collaborative manner on the customer needs.

Third, in considering the boundary-shattering aspect of digital, we notice that some of the core tenets of digital such as the Cloud, Cloud-native development, Microservices, Container-based deployments, Blockchain, Map Reduce, and Big Data processing are all decentralized or decentralization promoting constructs themselves. They promote the distribution of resources in a big way. Blockchain distributes used computing power among nodes that could be spread out geographically—in fact, decentralizing the capture of a transaction is a part of the value-appeal of Blockchain. Microservices' value proposition is based on writing code that can be built once and deployed everywhere—on any platform. Container-based deployments bundle the binary footprint of a microservice and its dependencies so that the microservice can be deployed anywhere and scaled up or down as guided by demand. Cloud, of course, takes away the codebase from the on-premise data centers of a company and outsources its location and management to the infrastructure of a cloud service provider—enabling the company to take a giant step towards decentralization. Cloud-native development through the use of the Twelve-factor app, Microservices, and product engineering creates an ability to scale by allowing the developed code to be deployed on any vendor's cloud platform, with minimal or no need for human involvement. Big Data processing uses data from all sources—distributed geographically—such as a company's transactional systems, B2B and IoT platforms, social media feeds, and even smart products or devices. It needs the processing horsepower of a decentralized platform such as the Cloud to be effective. Map Reduce is successful because it can distribute the required processing on commodity hardware.

On all these fronts—industry sectors, an organization's departments, and technology components—digital has shown a strong ability to remove boundaries. This trend will likely continue.

Thought number three: Automation devoid of intelligence will reach its non-differentiating destiny soon; the only way automation will survive as a competitive force is when it is intertwined with intelligence. It is the use of intelligence that will drive competitive differentiation.

In this book, we saw how automation or RPA provides value by having software mimic human or system actions so that other systems and workflows can be invoked in an automated way. We also saw how the value that RPA can provide has a ceiling: Only the structured, stable, and less error-prone business processes are the most suited for automation. This tends to restrict the value footprint of RPA to the most mundane, repetitive processes. Once such processes have been automated, the value realization from RPA reaches a saturation point. Now, augmenting RPA with intelligence can become the new source of value.

For instance, having Natural Language Processing—a form of Artificial Intelligence processing—scan the documents that are fed to an RPA deployment can increase the scope of services that RPA can provide. Arming RPA deployments with AI-ML informed decision-making would also fall in the same category of increasing the RPA's value scope. It stands to reason that as the footprint of AI-ML increases, the value realization from the RPA investments will not only increase but also help the company seek new frontiers of increasing competitive differentiation.

THOUGHT NUMBER FOUR: The dominance of the electronic D2C channel will probably overshoot. Customer experience through the physical channel will likely see a reemergence, albeit with a redefinition.

ON THE CHANNEL FRONT, it is likely that the dominance of the digital D2C channel will overshoot a bit. Some of the traffic to the physical store will see recovery. As the consumer gets more empowered through, for instance, continuing to have a voice on social media, having the ability to research product and brand options on her own, and then purchase and return using the channel of her choice, it is likely that while the eCommerce channel will continue to be the preferred channel for essential items, or for items when other constraints such as of time exist, the physical store will finder higher appeal as the channel of choice when the customer craves for an experience or some 'me-time'. The physical store of the future will be a part of a universe such as a digital mall where the customer will come to find a release from the burdens of modern life. The experience that this physical store universe will need to provide will transcend experience with only the brand. In this store, the customer would be offered a reprieve such as an ability to relax the mind, socialize with friends, spend time with oneself, explore items and things to do for hobbies and explore items for interest, and so on. This physical channel multiverse would prize customer experience to such an extent that it would go beyond the brand or its product lines and instead focus on what pleases and relaxes the customer. In so doing, this channel—in collaboration with like-minded brands—would gravitate the customers towards itself, over and over again. Of course, in the small moments, where the customer does intend to make a purchase, the Omni-channeled digital experience that the involved brands provide and that this mega-physical channel fosters is so seamless and powerful that it executes flawlessly, taking very little of customer's precious time to just make the purchase. Rewarded in this way, over time, the customer develops an intimacy with the brand or the basket of brands that are part of the channel multiverse and that have collaborated so imaginatively and beautifully to ease the burden on the customer's shoulders.

The self-less brands—the brands are selfless because they prize customer experience and place it at a higher priority than short-term gains—that focus in a laser-sharp manner on the customer's needs—explicit and hidden—will be rewarded by the digital world.

Thought number five: The traditional Management Consulting model will likely be upended by digital. The model will need to show on-the-ground value that can be created and show this value early.

Digital may upend Consulting in an irreversible way. The traditional management consulting model has been: spend weeks if not months doing analysis of company, industry, customer context, and so on, and use the findings to develop a sequence of recommended actions that the organization can take to improve its performance. The challenge with this approach is that the market dynamics fueled by a vibrant and ever-evolving technology front—and customer expectations—are too volatile, with the result that organizations need to continue to strengthen their abilities to experiment in a lean and fail-fast way. What this eliminates is the luxury of doing long analysis cycles that take even longer to show the potential face of their proposed results. Management consulting will need to respond to this need of the customer organizations it serves by not only showing the proposed value on the ground—with real customers, in real markets—but also showing this value early. The concept of MVPs—minimal viable products—that are developed quickly through close interaction with the customer is here to stay and is likely to find broader use with time. It is very likely that future management consulting models would make active use of MVPs to prove value.

THOUGHT NUMBER SIX: Digital will increase in strength as a voice for change, and because of its empowerment of all and its distribution effects, it likely will act as a major force for democracy and removal of inequality barriers.

AS A COROLLARY TO thoughts one and two, digital will increasingly act as a voice for and an enabler of change. In so doing, it will likely become a force for democracy and for removing the barriers of inequality. Evidence for this is already in place. In 2021, two Pulitzer Prize awards are noteworthy. The first, recognizing the "citizen journalism in the internet age," was awarded to a teenager who was not a journalist and who filmed the video of the killing of George Floyd[1]. The video went viral and set off protests worldwide against police brutality and led to measures to reform policing[2]. The second award went to a journalist who used satellite imagery and 3D architectural simulations to research and report on the detention camps targeting one minority group[3]. The journalist's work helped shed light—and alert other national governments—on a particular (and a major global player) nation's systemic oppression of the minority group.

The same digital consumer that today defines the expectations of a brand cares about and voices opinions on social and political issues—that quite often turn global with speed. This combination of motivated humans and technology will create a championship for issues as varied as climate change, localized suppression or inequity issues, and gender equality.

1. Ians. "Indian-Origin Journo Megha Rajagopalan Wins Pulitzer for Exposing China's Muslim Detention Camps." *OnManorama*, 12 June 2021, www.onmanorama.com/news/world/2021/06/12/indian-origin-journo-megha-rajagopalan-wins-pulitzer-for-exposing-china-muslim-detention-camps.html.
2. Ibid.
3. Ibid.

The empowered consumer and her voice will also likely drive the mitigation of existing regulatory deficit vis-a-vis the digital giants—the social media leaders; Google and; Apple. It will force the giants to show higher deference for customer information and her rights to privacy. The consumer will win the war on rights to privacy—it is only a matter of time. The free-market forces ironically are the ones that will make this happen, as customers will reward and gravitate to those companies that protect or even champion their right to privacy.

Thought number seven: Artificial Intelligence will continue to be the game-changer for the foreseeable future.

Computers, even their oldest incarnations—the computing devices such as calculators—have always embedded intelligence. The oldest calculator could add two numbers or multiply them and return the correct result. It could do so because the calculating formulae—the intelligence—were hardcoded into them. Years later, computers took on the operations of entire business processes. They—instead of humans—could run the entire business process workflows. The business rules—the new intelligence—were coded into the computers through software using the if-then logic. In both the cases—that of the calculating device or the business process modeling computing systems—though, the intelligence that was embedded was static. It could not—or was not readily amenable to—change with time. In the case of the calculating device, it literally was the case. If, by some random stroke, 2 + 2 did not equal four, all the bought calculators would suddenly become useless. In the case of the systems, the impacts are not so drastic. New business rules—the new intelligence—could be incorporated, but that would require changes to the existing software—new software features would need to be designed, built, tested, and deployed into production. The Y2K issue at the turn of the century a couple of decades ago was an example. Because the date format had been restrictively designed, the issue presented the risk of crashing computer operations worldwide when the clock struck midnight.

In this sense, Artificial Intelligence and Machine Learning are game-changer. AI-ML is different in how it treats intelligence. It learns by experience, can train itself on real-world data, and optimization of its accuracy performance—that can mostly be automated—allows it to get better with time. That is, for AI-ML, potentially sky is the limit. As long as the ever-hungry model can be fed with data, it will deliver, and even get better—and more powerful?—with time. Potentially one day, the AI-ML will go from mere-mimicry of human actions to mockery—of the inadequacy of the human.

When all digital trends have had their day in the sun, there, in all likelihood, will be one final reigning emperor, and that will be Artificial Intelligence.

Index

Made in the USA
Monee, IL
04 November 2026

67384742R00125